International

Directory

of

Hemp

Products

and

Suppliers

compiled and edited by
James Berry

The
Message Company

LIBRARY OF CONGRESS CATALOG CARD NUMBER: 96-094235
ISBN: 1-57282-005-5

Cover design by Janice St. Marie
Book Design by James Berry

677   Publisher's Cataloging in Publication
BER   Berry, James William
         International directory of hemp products and suppliers /
         compiled and edited by James Berry. - Santa Fe, N.M. : The
         Message Company, ©1996.

         144 p.
         Includes index.

         Summary: A complete directory of the burgeoning hemp
         industry, listing manufacturers, growers, importers, distributors,
         retailers and more, plus indexes by country and state, and by
         product.

         ISBN 1-57282-005-5

         1. Hemp - Directories 2. Hemp industry - Directories
         3. Plant fiber industry - Directories I. Berry, James William

                      677'.12_dc20

Published and distributed by:
The Message Company
4 Camino Azul
Santa Fe, NM 87505
505-474-0998

Printed in Canada
Printed on acid-free recycled paper using soy ink

TABLE OF CONTENTS

The World History of Commercial Hemp

Hemp Travels The World:

1.	3000	BC	China
2.	1000	BC	India
3.	500	BC	Africa-Asia
4.	500	AD	Europe
5.	1495	N	America
6.	1545	S	America
7.	1992		Australia
8.	1993		England
9.	1994		Canada

Hemp has a history that goes back to pre-historic time. Hemp has been found in tombs dating back to 3,000 BC [1].

The use of hemp can be documented as far back as 2700 BC in ancient Chinese writings.

These writings tell us that hemp was used by the Chinese for a variety of uses. These included fiber, oil, and as a medicine. By 450 BC, history tells us that hemp was being cultivated in the Mid-East region. From Afghanistan to Egypt, hemp was cultivated for its fiber and drugs. It appears that hemp was first introduced into Europe between 500-1000 AD. It is known that hemp was in wide cultivation in Europe by the sixteenth century. It was cultivated for its fiber and its seed.

The seed was cooked with barley and other grains and eaten. In 1537, Dioscorides called the plant Cannabis sativa, the scientific name that stands today as its true name. He noted its use in "the stoutest cords" and also its medicinal properties [2].

Hemp was introduced into Chile about 1545 [3] where it was grown for fiber. Hemp was introduced in New England soon after the Puritan immigrants settled, noting that it grew "twice so high."[4] In Virginia, the early legislature passed many acts to promote the hemp industry. Before the revolution, hemp flourished in the area around Lancaster, Pennsylvania. Hemp was first grown in Kentucky in 1775 [5]. In 1802, two extensive rope walks were built in Lexington Kentucky. There was also announced a machine that could break "eight thousand weight of hemp per day"[6], a huge quantity for the time. Hemp spread to other states including Missouri by 1835, Illinois by 1875, Minnesota by 1880[7], Nebraska by 1887, California by 1912[8], Wisconsin and Iowa by the early 1920's.

The cultivation of hemp was stalled by federal action in 1937 by the imposition of heavy regulations on producers known as the Marijuana Tax Act. By 1940, the US government reduced the regulations so production could take place during World War II. After World War II, with the regulations reinstated, cultivation declined until the 1950's when the last commercial crop was grown in The United States.

Footnotes: 1 *Yearbook of the Dept. of Agriculture*, L. Dewey, Pg. 296, 1913.
2 Dioscorides, *Medica Materia*, li bri sex, pg. 147, 1537.
3 Husbands, Jose D., *US Dept. of Agriculture, Bureau of Plant Industry, Bulletin #153*, pg. 42, 1909.
4 *Yearbook of the Dept. of Agriculture*, L. Dewey, pg. 291, 1913.
5 Moore, Brent, *A Study of the Past, the Present and Future of the Hemp Industry in Kentucky*, pg. 16, 1905.
6 Michaux, Andre, *Travels to the West of the Alleghenies*, pg. 152, 1805.
7 Schoenrock Ruth, *Hemp in Minnesota During the War Time Emergency*, Pg. 15, 1996.
8 *Yearbook of the Dept. of Agriculture*, L. Dewey, pg. 293, 1913.

SIGNING THE DECLARATION OF INDEPENDENCE

The Earth's Premier Renewable Resource

by John Roulac

While forests diminish worldwide, prices and demand for fiber are skyrocketing. A plant cultivated by America's founding fathers may be the solution to our growing fiber shortage. Imagine a crop more versatile than the soybean, the cotton plant, and the Douglas fir put together...one that grows like Jack's beanstalk with minimal tending. There is such a crop: industrial hemp. Hemp was once indispensable to world commerce. The 1913 *Yearbook of the US Dept. of Agriculture* called hemp "the oldest cultivated fiber plant," mentioned how the crop improves the land, and said that it yields "one of the most durable fibers of commerce."

Then, in 1937, fiber hemp fell victim to the anti-drug sentiment of the times when the US Congress passed the Marijuana Tax Act. The intent of the law was to prohibit the use of marijuana, yet it created so much red tape that the production of industrial hemp became nearly impossible. Several farmers continued to grow hemp under license in the state of Wisconsin, until they let their permits lapse in the 1950s. The fact is, hemp grown for fiber, whether by George Washington in

1790 or by English farmers in 1995, has never contained psychoactive qualities.

FARMING HEMP

Industrial hemp gives farmers a crop that produces a high-quality fiber without the use of any herbicides or pesticides. The June 1994 issue of *National Geographic* stated that "in California alone some 6,000 tons of pesticides and defoliants are used on cotton in a single year." Since hemp plants grow 6 to 16 feet tall in 70 to 110 days, farmers of large and small acreage alike can shade out weeds and thus eliminate the use of costly herbicides.

The wide variety of products that can be made from hemp through sustainable agriculture may help to fulfill the prophecy made by the great American scientist, George Washington Carver, when he many years ago said, "I believe that the great Creator has put ores and oil on this earth to give us a breathing spell. As we exhaust them, we must be prepared to fall back on our farms, which is God's true storehouse and can never be exhausted. We can learn to synthesize material for every human need from things that grow."

With the tobacco industry in decline, there is a strong interest among tobacco farmers in the cultivation of hemp. The

Kentucky Hemp Growers Cooperative Association of Lexington, Kentucky, a 53-year-old organization comprised of 60 farmers, is currently in negotiation with a Fortune 500 paper manufacturer to use hemp as a fiber supplement. A recent poll conducted by Kentucky University reported that 76% of Kentuckians favor licensing industrial hemp in the state of Kentucky. Colorado State Senator Lloyd Casey plans to introduce in 1996 the Colorado Hemp Production Act that would permit farming of industrial hemp varieties. Positive stories on hemp have been published by the *New York Times*, the *Los Angeles Times,* the *Boston Globe,* and the *Washington Post* .

COMMERCIAL POTENTIAL

Hemp's versatility yields a range of material— fiber from the stock, and seed, seed oil, and seed cake (crushed seed)— that can be processed into thousands of valuable products. In 1938, *Popular Mechanics* magazine stated, "Hemp is the standard fiber of the world. It has great tensile strength and durability ...(and is) used to produce more than 25,000 products, ranging from dynamite to cellophane."

Annual sales of hemp goods in North America, which totaled only a few million dollars in 1993, are projected to approach $50 million in 1995, and $500 million by the year 2000. Adidas is jumping on the hemp bandwagon with a line of hemp sneakers, and fashion companies such as Ralph Lauren are finding use for hemp's long-lasting, high-quality fabrics. Industrial hemp fibers are currently being tested as a substitute for fiberglass in advanced composites for electric vehicles, as well as for non-woven applications in manufacturing rugs.

COMPOSITES

The products known as composites, including paneling, medium-density fiberboard, trusses, and support beams, comprise the fastest-growing segment of the wood-products industry. Washington State University's preeminent Wood Composite Laboratory has tested hemp for use in medium-density fiberboard, and lab results show that hemp is up to twice as strong as wood. According to lab director Tom Maloney, "The use of hemp fiber in multidensity fiberboard and other composites looks very promising." Unless northwest towns are provided with a new economical fiber source such as hemp, thousands of mill jobs will be lost in the coming years.

FREE TRADE BLOCKED

Our economic trading partners, including Britain, China, France, and Holland, are growing and processing hemp and exporting millions of dollars of hemp goods such as foods, footwear, paper, textiles, and cosmetics to the United States. US farmers and manufacturers are restricted from such activities, as the US Government has not granted any permits for large-scale hemp farming in over 40 years. Our neighbor to the north, Canada, issued permits for seven hemp farms in 1995 and is gearing up for future production to be exported into the US markets. This is an ironic twist for the United States of America, upholder of global free trade and signer of the North American Free Trade Agreement and the General Agreement on Trade and Tariff. Yes, hemp is mentioned in NAFTA Article 708 53.02. The US Government's current policy is increasing the American trade deficit while infringing on the economic rights of American farmers and business people.

THE NEXT STEP

The December 17, 1994 Canadian Government Agriculture Department Bulletin, printed on hemp paper, stated, "A reawakening to hemp's industrial potential is being observed worldwide." As more organizations come to the conclusion that the present use of oil, timber, and cotton is not sustainable, hemp's role as a premier earth-friendly raw material will only increase. Many agricultural and manufacturing insiders are now predicting that hemp will be the industrial crop of the 21st century, rivaling corn and soybeans in the number of acres planted.

In the coming years, free market traders will continue to jump over the US "Hemp Wall" to deliver imported hemp products. Purchasers of hemp products are voting with their dollars, and the race is on to market hemp goods to this expanding group of buyers. American farmers and manufacturers cannot afford to be left behind.

Hemp Paper

by John R. Stahl

Hemp has been the mainstay of the paper industry for over 2000 years. The earliest fragments of paper have been found in China, and dated at around 3000 BC. These early sheets have been analyzed and shown to be composed of hemp, cannabis sativa (see *"On the Origin of Paper Making in the Light of Scientific Research on Recent Archeological Discoveries,"* by Dr. Jixing Pan, 1986). Since that date, other materials have been used (ramie, flax, cotton, etc.), but it has not been until the mid nineteenth century and the discovery that a variety of paper can be made from wood chips (albeit a very poor paper which has a maximum useful life of about 50 years, compared with about a thousand years for a well made sheet of hemp paper) that hemp ceased to be the principle fiber employed in paper.

There are good reasons for this. In the first place, hemp is one of the longest and strongest natural fibers known to man. In the second place, it is one of the most economical plants to grow, yielding an enormous harvest of high quality fiber in a single growing season.

However, the reason for the transition to wood chips for paper pulp was the perception that wood was a basically free and inexhaustible resource. We now know differently, of course: the price of wood chips (and the paper made from it) has been going steadily up, and those in the industry know all too well that the resource is very far from inexhaustible!

So why is hemp paper so slow to join the renaissance of industrial hemp? The reason is that the commercial pulp and paper industry is heavily invested in wood pulp technology, and this technology is not suitable for the production of pulp from hemp. At the present time, just about every scrap of hemp paper on the market (with the exception of the hemp paper coming from China) is produced from pulp produced from one mill in Spain, and it is expensive. This includes all the European, English, and American sources. There may be some very small scale exceptions to this, and there will be further diversity very soon. I know there is a major mill in Germany coming on line which is planning to prepare its own pulp, and I am personally coordinating the funding for a new pulping technology coming from the Ukraine, which should be available within a year or two.

I make paper by hand, and I have been experimenting with hemp fiber for about 5 years now. At first, I used anonymous donations of hemp stalks grown here in Northern California, but I have also used bast fiber hemp (the long outer fiber--the material used for textiles and fine paper) imported from China. My experiences were so promising that I obtained several bales of partially processed hemp pulp from Spain, and began using hemp fiber as the base for most of the hand made papers that we make. Using the local material is a lot of fun, but it is enormously labor intensive, because the material comes to me in very random form, all sticks and short branches, since the material has been cultivated for tops rather than long fiber. (Even though it is "sinsemilla," it is comparable in its growth habits to a commercial hemp crop grown for seed as opposed to a textile fiber crop.) Consequently, I have applied for a permit to cultivate fiber hemp for paper. The DEA has finally (after several years of negotiation) agreed that I am in full compliance with their regulations (8 foot chain link fence, barbed wire top, locked gate, a safe to store viable seed, floodlights, alarms, and a 24 hour guard!) and is prepared to grant me a permit as soon as I can work out an exemption from California's laws against cultivation of cannabis. Hopefully, this will be obtained in my lifetime.

In the meantime, we go on experimenting and making different kinds of hemp paper. Our early paper made from the local hemp utilized the whole stalks, because of the difficulty of separating the bast fiber. Our paper made from the Spanish hemp is better quality since it is made from just the bast fiber. Currently our most popular hemp paper is a blend containing about 10% local hemp and the balance Spanish hemp. We are also working on a blend of 50-50, which we expect to be very fine. This year we have begun production of a very limited supply of pure local bast fiber hemp paper. The labor for this is so monumental that we couldn't offer to sell this paper at any price. However, we are making enough for inclusion in a book featuring hand made paper made from scratch from indigenous plants. We are expecting to have enough pulp left over to make a very limited batch of hand made rolling papers from this local bast fiber hemp. Again, we will not be able to sell this; it will just be given to very special friends and customers. We are, however, (due to popular demand) beginning to make Pure Hemp rolling papers from the Spanish bast fiber hemp material, and these we are able to offer to the trade.

My interest in hemp paper extends to the commercial paper industry, since it troubles me to see our forests devoured to produce junk mail. Hemp holds out the promise of being able to produce a superior paper product at a lower price than the wood pulp competition. The problem is that in order to realize this dream, there is a considerable technological hurdle to be

18

overcome before we are in a position to compete directly with the established paper industry. In addition, once we have developed a successful process for making hemp based paper, we will need to set up production on a $700,000,000 production facility that can produce at least 1000 tons of paper a day, in order to compete on a level playing field!

Much research remains to be done, but the current idea, as I now see it, is to cultivate hemp primarily as a seed crop since there is a limitless demand for the very valuable seed, both for industrial applications and for human nutrition. Once again, as with hemp grown for fiber, one of the principle attractions of a seed crop to the farmer is the enormous harvest produced every season. Then, once the seed has been harvested, the entire remaining stalk material, both bast fiber and hurds (woody core) can be chipped up and processed into pulp for paper making. Since the core contains about 70% of the plant, it is essential to include it in any commercial paper making operation. (All currently available hemp paper uses only the bast fiber.) It is true that the hurds are composed of very short fibers, but they are high in cellulose, and, when combined with the longer bast fibers, produce excellent pulp for paper making. Furthermore, in order to keep the cost of production down, we advocate the use of agricultural and industrial waste material, such as wheat straw and cotton scraps.

The new pulping design from the Ukraine which we are developing is a closed system which will produce no toxic effluents (and will, on the contrary, leave a high nitrogen fertilizer as its only by-product), and will be able to process a wide variety of source material, allowing the mill to operate as a real recycling center for cellulose material from any source. Altogether, the future looks very bright for the return of hemp to its rightful place in the sun.

THE REDISCOVERY OF HEMP SEED OIL

by Ron Lampi

The hemp plant is slowly regaining its once well-deserved reputation as an important, extremely useful farm crop. Up until the 1930's, fields of hemp were a common sight on many farms in our country. It has provided humanity with valuable resources of versatile fiber, edible seeds and oil, and medicines for thousands of years. Hemp fiber is strong and durable; it was used traditionally for making sails, canvas, twines, ropes, cordage, cloth, clothing, and more recently, plastics. Hemp seeds are an important food source in some parts of the world; they contain high quality protein and fat in the form of a highly nutritional oil.

Hemp seed oil is the new food oil on the market these days. It is a nutritionally superior oil that has been called nature's perfect blend of the essential fatty acids Omega-3 and Omega-6. Essential fatty acids (EFAs) are nutrients that are required in our diet, since our bodies cannot produce them. They are critical in all cell functions and body systems, and have proven effective in curbing and preventing many disease conditions. They help us to digest fats. An essential fatty acid deficiency can often result in high cholesterol levels.

Many people take oils as a supplement to their diet. Flax oil has been well-known for years. Hemp oil, however, is a wider spectrum nutritional oil than flax. Hemp oil also contains Gamma-linolenic acid (GLA), an important fatty acid that promotes healthy skin, hair, and nails, and helps in reducing inflammation. The usual supplemental source of GLA is evening primrose oil. But hemp oil has it all--a rich blend of Omega-3 and Omega-6 EFAs, GLA, in addition to Omega-9, palmitic acid and other important unsaturated fatty acids. Of all foods that contain fatty acids, hemp oil is the only one that contains both the Omega-EFAs and Gamma-FAs. And because of its high fatty acid profile, it is one of the most digestible and assimilable of foods.

Unlike other oils that are nutritionally good for you, hemp oil has a pleasant, clean, nutty flavor. It can be used in a wide variety of ways, easily replacing the oil you have been using. Try it in your favorite recipes for dressings, dips, pesto, and spreads. You can use it as a topping over pasta, rice, and potatoes. Try it on whole grain breads, crackers, and bagels. For a simple idea, mix hemp oil with garlic, lemon, and sea kelp or red dulce and pour it on breads, pasta, salads, or potatoes. It can be used as a butter substitute, for example, in popcorn. Hemp seed oil is as versatile in restaurants as it is in the home. A number of restaurants in Hawaii and California have featured it on menus and at banquets; successful gourmet events have highlighted dishes using hemp oil.

Hemp oil has many other uses in addition: in natural health care and first aid, massage, soap-making, and wood preservation. Refined hemp oil is excellent for manufacturing purposes, including making soap, cosmetics, industrial lubricants, and machinery oils. The traditional uses of hemp are staggering. Creative minds, rediscovering it, will find even more.

Manufacturers who make lip balms using hemp seed oil are The Merry Hempsters, Hemp Essentials, and Sue's Amazing Lip Balm. Artha is a fine hemp soap manufacturer, and Linda Kammins, an Aromatherapist in Los Angeles, makes cosmetics using hemp oil. In foods, we have Dr. Alan Brady in Santa Cruz who makes a hot and spicy hemp oil dressing, hemp ice cream, and hemp halvah. Hungry Bear Hemp Foods in Oregon makes a hemp seed snack food treat called Seedy Sweeties. We at Herbal Products & Development are collecting hemp recipes for a book-in-progress we call The Hemp Uncook Book.

Here are two recipes to experiment with hemp seed oil:

Pesto

1/3 cup hemp seed oil
1 bunch basil
1/3 cup pine nuts or walnuts
1 cup parsley with beet greens
filtered water (optional)
2 cloves garlic
small lemon

Blend together to taste. Note: add water or hemp seed oil as needed to keep paste consistency. Serve & enjoy!

Hemp Land & Sea Spread

(For bagels, whole grain breads & crackers. Good for potlucks.)

1 oz. hemp seed oil
medium to spicy hot sauce (dash)
1/2 oz. vegetable powder or Dr.
      Bronner's seasoning mix
Red dulce (to taste)
1 oz. flax meal or oil
pumpkin seeds (diced)
whole grain mustard (knifeful)
1/4 oz. Dr. Bragg's liquid aminos
small lemon

1 package tofu
small red onion (finely diced)
ground sesame seeds (to taste)
1/8 tsp. caraway seeds

Mix with mixed sprouts on whatever you choose!

Herbal Products & Development offers extra-virgin, unrefined hemp seed oil from the South American Andes. It is made from non-sterilized seeds that are organically grown and cold-pressed to assure the finest quality. Being very concerned that only high quality hemp oil reaches our customers, we immediately store the oil in a frozen state until it gets air-freighted to its destination. During the flight, and after arrival in the US, it is maintained in a near-frozen state. Along with these precautions, we seal all drums of oil with nitrogen, and use natural rosemary antioxidant to retard spoilage.

Strength and Elongation of Various Fibers [1]:

Fiber:	Denier	Tenacity	Break Length [2]	Tensile Strength	Elongation %
Ramie (Yuenkang)	5.42	7.34	66.0	99.0	4.6
Ramie (Kingkiang)	5.88	6.73	60.6	91.0	4.2
Jute	19.55	3.19	28.7	41.4	1.4
Flax	16.5	5.82	52.4	76.5	1.6
True Hemp	12.75	6.26	56.4	83.5	2.0
Cotton, Am	2.10	2.74	24.6	36.9	9.8
Silk	1.27	3.76	34.0	46.3	16.3
Wool, NZ	15.00	2.06	18.5	24.6	40.9

Footnotes:
1. Handbook of Textile Fibers, page 135.
2. Length in Km - a fibers' break length is a measurement that shows how long a fiber would have to be to break under its own weight. Usually measured in kilometers. Used with Permission of The Institute for Hemp.

Dimensions of Fibers & Cells of Vegetable Fibers[1]:

Fiber	Diameter, m m Min/Max/Avg			Break Length[3]	Ultimate Strain%	Work of Rupture	Initial Elast
Nettle	0.020	0.070	0.042				
Ramie	0.017	0.064	0.040	32-67	2-7		
Sunn Hemp	0.013	0.061	0.031				
Hemp	0.013	0.041	0.025	38-62	2-4	0.6-0.9	180
Manila Hemp	0.010	0.032	0.024	32-69	2-4.5	0.6	
Kenaf	0.013	0.034	0.023				
Sansevieria	0.013	0.040	0.022				
Sisal	0.007	0.047	0.021	30-45	2-3	0.7-0.8	2500-2600
Mauritus	0.015	0.024					
Coir	0.010	0.024	0.020				
Cotton	0.014	0.021	0.019				
Flax	0.008	0.031	0.019	24-70	2-3	0.9	1800-2000
Kapoc	0.010	0.030	0.018	16-30	1.2	0.1	1300
Jute	0.005	0.025	0.018	27-53	1.5	0.3	1700-1800

Footnotes: 1. Handbook of Textile Fibers, Harris Research Lab.
3. Length in Km - a fibers' break length is a measurement that shows how long a fiber would have to be to break under its own weight. Usually measured in Kilometers. Used by permission of The Institute of Hemp.

Dimensions of Fibers & Cells of Vegetable Fibers[1]:

Fiber	Ultimate[2] Fiber Length, mm			Fiber Strands	
	Min	Max	Avg	Range/cm	Width/mm
Ramie	60	250	120	10-180	0.06-9.04
Nettle	4	70	38		0.04-0.62
Flax	8	69	32	20-140	
True Hemp	5	55	25	100-300	
Cotton	10	50	25	1.5-5.6	0.012-0.0025
Kapoc	15	30	19		
Sunn Hemp	2	11	7		
Manila Hemp	2	12	6	180-340	0.01-0.28
Phormium	2	11	6		
Pineapple Fiber	2	10	5.5		
Kenaf	2	11	3.3		
Sisal	0.8	7.5	3	75-120	0.01-0.28
Jute	0.75	6	2.5	150-360	0.03-0.14

Footnotes: 1. Handbook of Textile Fibers, Harris Research Lab. 2. Ultimate means Individual Fibers. Used by permission of The Institute of Hemp.

28

Comparison of Physical Properties of True and Sunn Hemp Fibers

Comparison Breaking Strength of True Hemp and Sunn Hemp Cordage after Exposure to Fresh and Salt Water[1]

| | True Hemp | | Sunn Hemp | |
| | Time (days) | Break Strength[2] | Time (days) | Break Strength[2] |
Condition				
Dry		126.0		91.8
Salt	30	96.4	30	66.1
	60	50.5	60	35.9
	90	7.1	90	3.2
Fresh	40	103.0	40	83.0
	80	101.9	80	77.1
	160	102.0	160	72.1
	282	92.2	282	70.3

Footnotes: 1 Textile Age, 8:70, 1994.
2 In Pounds.
Used by permission of The Institute for Hemp

Alphabetical

Directory

of

Hemp

Suppliers

2000 BC
Dave Ratcliffe
8260 Melrose Ave., Los Angeles, CA 90046
USA
213-782-0760   Fax: 310-289-7905
Retailer
Specializing in hemp products and accessories.
Products: Men and women's clothing, handbags, wallets, accessories, hats, packs, outdoor gear, shoes, sandals, sheets, pillows, towels, wash cloths, fabric, rope, twine, printing paper, rolling papers, soaps, massage and body oils, salves, lip balms, seeds (roasted), cookies, cookie mixes, pancake mixes, vitamins, supplements, hacky sacks, books, videos, consumer periodicals.

AdventureSmiths
Sue Smith
PO Box 50333, Eugene, OR, 97405
USA
541-343-9924   Fax: 541-343-6103
Manufacturer, Retailer
Handmade hemp clothing and accessories.
Products: Men and women's clothing, handbags, wallets, accessories, hats, packs, outdoor gear, fabric, hacky sacks.

AH Hemp
Todd Brown
1500 Park Ave., #A304, Emeryville, CA 94608
USA
510-595-9225   Fax: 510-595-9694

E-mail: ahhemp@aol.com
Manufacturer, Wholesaler.
Boutique style traditional and classic hemp clothing and accessories.
Products: Men and women's clothing, handbags, wallets, accessories, hats, packs, outdoor gear, towels, wash cloths, soaps.

Alaska Green Goods
3535 College Rd., Fairbanks, AK, 99709
USA
907-452-4426 Fax: 907-452-4136
Toll-Free: 800-770-4426 (within Alaska)
Retailer
Environmental products retail store featuring hemp products, organic cotton clothing, natural personal care items, gifts.
Products: Men and women's clothing, handbags, wallets, accessories, hats, rope, twine, specialty papers, soaps.

All Around the World Hemp
Jeffrey Stonehill
PO Box 335, Lopez, WA, 98261
USA
Toll-free: 800-449-4945 Fax: 360-468-3374
Distributor, Manufacturer
Researches products from hemp oil and manufactures fiber paper and hemp rugs. Also lectures at businesses and schools.
Products: Fabric, printing paper, rope, twine, soaps, lip balms, food grade oil (bulk). Currently researching paints and varnishes,

33

lubricants, fuel, fiberboard and cement blocks for building.

All Points East
Darin Novak
PO Box 221776, Carmel, CA 93922
USA
408-655-HEMP   Fax: 408-655-HEMP
Distributor
Introduces hemp products to the environmental market using all natural vegetable dyes.
Products: Men and women's clothing, handbags, wallets, accessories, hats, printing paper.

Alma Rosa N.V. Belgium
Gunter
Belgium
32-328-14595   Fax: 32-328-14596
Distributor
Distributes Canna Bliss hemp seed oil body care products and Alma Rosa pure hemp and hemp/recycled paper for printing.
Products: Printing paper, soaps, shampoos, salves, lip balms.

American Hemp Mercantile
Ken Friedman
502 Second Ave. #1323, Seattle, WA, 98104
USA
206-340-0124   Fax: 206-340-1086
Toll-free: 800-469-4367

Retailer, Manufacturer, Distributor, Importer
Imports hemp twine, fabric, and paper from
Europe, and manufactures hemp apparel and
accessories for sale through wholesale
accounts and its two retail stores.
Products: Men and women's clothing,
handbags, wallets, accessories, hats, packs,
outdoor gear, fabric, rope, twine, nets, sails,
marine products, canvas, tarps, printing
paper, soaps, shampoos, massage and body
oils, salves, lip balms, cosmetics, granola,
seeds (raw), food grade oil (bulk), hacky
sacks, coffee filters, books and videos.

AMPT
Bryan Sturgill
PO Box 1743, Athens, GA, 30603
USA
706-369-7646 Fax: 706-369-7641
Retailer, Distributor, Manufacturer
Manufacturers and wholesales hemp jeans.
Supplies the action sports market with quality
garments and water based inks.
Products: Men's and women's clothing.

Artha
Allysyn Kiplinger
PO Box 20154, Oakland, CA, 94620-0154
USA
510-420-0696 Fax: 510-420-0696
Manufacturer

Handmade vegetarian soap using hemp oil and flour. Soap is scented with 100% essential plant oils. Packaged in hemp paper.
Product: S o a p

Artisan   Weavers
Netaka J. White
PO Box 76, East Middlebury, VT, 05740
USA
802-388-6856   Fax: 802-388-6856
Manufacturer
Designs and produces American hand-crafted travel bags and accessories made from hemp, hand-woven wool and recycled fibers. Wholesale and Retail.
Products: Handbags, wallets, accessories, packs, outdoor gear.

ASA   Aware   Products
Jeff Joseph
625 SW 10th Ave., #340, Portland, OR 97205
USA
503-235-3583   Fax: 503-235-3583
Toll-Free: 800-6-SATIVA
Retailer, Manufacturer, Contract fabric and garment dyeing.
Embroiders hemp patches and manufactures day packs, fanny packs, hats and other products.
Products: Handbags, wallets, accessories, outdoor gear, backpacks, hats, fabric.

Australian Hemp Products
Grant Steggles
104 Darby St., Newcastle 2300
Australia
61-49-265371   Fax: 61-49-294336
Retailer, Distributor, Manufacturer, Importer
Largest Australian hemp business.   First to
have hemp t-shirts in the world.
Products: Men and women's clothing,
handbags, wallets, accessories, hats, packs,
outdoor gear, shoes, sandals, fabric, rope,
twine, canvas and tarps, technical papers,
rolling papers, soaps, shampoos, massage and
body oils, salves, lip balms, cosmetics, molded
parts and plastics.

The Boulder Hemp Company (AKA One
Brown Mouse)
Michael Perkins
PO Box 1794, Nederland, CO, 80466
USA
303-443-7875
Distributor, Manufacturer
Manufactures and distributes hemp seed flour
foods.   Boulder County Health Department
approved.
Products: Cookies, dry food mixes (pancake,
cookie and all purpose baking mix) using
hemp seed flour, plus Hempburger mix.

British Hemp Stores
Tim Barford
76 Colston St., Bristol BSI 5BR
England
0117-9298371 Fax: 0117-9238326
Retailer, Distributor, Manufacturer, Importer
Retail outlet. Manufactures own label
(Allesandro Dasosa), distributes prime USA
and UK labels and imports hemp textiles.
Products: Men, women and children's
clothing, handbags, wallets, accessories, packs,
outdoor gear, fabric, rope, twine, printing
paper, soaps, shampoos, massage and body
oils, salves, lip balms, cosmetics, seeds (raw),
salad oils, trade magazines, books and videos.

Business Alliance for Commerce in Hemp
Chris Conrad
PO Box 1716, El Cerrito, CA 94530
USA
510-215-8326 Fax: 510-215-8326
Association.
Products: Publication: *Lifeline to the Future.*

Cannabest
Lee Neel, Lyle Nishwonger
PO Box 12960 San Luis Obispo, CA 93406
USA
805-543-4213 Fax: 805-544-4076
Toll-free: 800-277-0510
Retailer, Distributor
Full color catalog printed on hemp paper
available.

Products: Men and women's clothing, handbags, wallets, accessories, hats, packs, outdoor gear, shoes, sandals, sheets, pillows, futons, fabric, rope, twine, canvas, tarps, printing paper, specialty papers, rolling papers, newsprint, soaps, shampoos, massage and body oils, salves, lip balms, seeds (raw and roasted), food grade oil (bulk), salad oils, dressings, hacky sacks, birdseed, trade magazines, books and videos.

Cannabis Clothes
Candi Penn
PO Box 1167, Occidental, CA, 95465
USA
707-874-1104   Fax: 707-874-1104
E-mail: hempstrs@wco.com
Retailer
Hooded jackets (100% hemp shell, cotton flannel linings), hemp/silk skirts (3 lengths).
Products: Men and women's clothing.

Cannabiz Company
Dave Barsky
PO Box 22823, Santa Fe, NM 87502
USA
505-470-3400
Retailer, Distributor, Importer
Hand-woven hemp products.   Constantly changing product line.
Products: Men and women's clothing, handbags, wallets, accessories, hats, fabric, hacky sacks.

Canvas Hemp Company (CHC)
Tim Hines
PO Box 3705, Dana Point, CA 92629
USA
714-240-7572
Manufacturer
Inventors of the original hemp surfboard.
Products: Handbags, wallets, accessories, shoes, sandals, surfboards.

CEL BT
Ursula Harrach
3950 Sarospatak PF 39
Hungary
Fax: 36-36-4111-467
Manufacturer.
Hemp and mixed (hemp/straw) paper sized 65 x 92 cm, weights (80, 100, 240 G/M$_2$).
Products: Printing paper.

Chanvre en Ville
Larry Duprey
3418A Ave. du Parc, Montreal, Quebec
H2X 2H5
Canada
514-845-4993 Fax: 514-487-0651
Retail Store
Products: Men and women's clothing, handbags, wallets, accessories, hats, packs, outdoor gear, shoes, sandals, fabric, rope, twine, printing paper, technical papers, soaps,

massage and body oils, salves, lip balms, seeds (roasted and raw), hacky sacks.

Community Market
Melissa Minton
1899 Mendorino Ave., Santa Rosa, CA 95401
USA
707-546-1806   Fax: 707-546-6555
Retailer
Natural Foods and Products Store.
Products: Handbags, wallets, accessories, hats, rope, twine, printing paper, massage and body oils, salves, lip balms, seeds (roasted and raw), food grade oil (bulk), trade magazines, consumer periodicals.

Cotton and Willow
Monica Fleming, Rob Marchak
313A 19th St., NW Calgary T2N 2J2
Canada
403-283-8946   Fax: 403-283-8946
Importer
Canada's first importer of high quality hemp textiles.
Products: Fabric, futon covers.

Creative Expressions
Chris Conrad
PO Box 1716, El Cerrito, CA, 94530
USA
510-215-8326   Fax: 510-215-8326
Toll-free: 800-HEMP-MAN
Publisher, Consultants
Book: *Hemp: Lifeline to the Future*
Products: Books and videos.

Crop Circle Clothing
Christian Weber
2442 NW Market St. #50, Seattle, WA 98107
USA
206-726-3990   Fax: 206-985-2798
Toll-free: 800-618-4367
E-mail: ch20@aol.com
Web: http://www.cropcircle.com
Distributor, Manufacturer, Importer, Sourcing
Consultants.
Clothing and accessories made of 100% hemp
in the widest range of bright and vivid
colorfast colors.
Products: Men and women's clothing,
handbags, wallets, accessories, hats, packs,
outdoor gear, shoes, sandals, fabric, rope,
twine, nets, sails, marine products, canvas,
tarps, hacky sacks.

Crucial Creations
Denny Finneran
4550 S. 12th Ave. #111, Tucson, AZ 85714
USA
Toll Free: 800-HEMP-4-US
Fax: 520-682-7804
Manufacturer
High fashion clothing from 100% hemp.
Products: Men, women and baby's clothing, diapers, handbags, wallets, accessories.

Cultural Repercussions
Dinair Wolf
PO Box 1301, Bisbee, AZ, 85603
USA
602-230-5242
Manufacturer
Designer and manufacturer of garments, hats and bags for alternative lifestyles.
Products: Men, women and children's clothing, handbags, wallets, accessories, hats.

Danforth International
Frank Riccio, Jr.
3156 Rt. 88, Point Pleasant, NJ 08792
USA
908-892-4452   Fax: 908-892-1421
Importer
One of the world's largest producers and suppliers of non-wood fibers and pulps to the specialty paper, textiles, non-wovens, thermo-plastics and composite materials industries.

Products: Raw fibers, fabric, rope, twine, paper pulp, fiber for paper, yarn, fibers for non-wovens.

Deep See
Dolphin Dreaming
501 N. 36th St. #236, Seattle, WA 98103
USA
Toll-free: 800-436-7783
Distributor
Distributor of a wide variety of hemp products.
Products: Handbags, wallets, accessories, hats, shoes, sandals, fabric, rope, twine, rolling papers, soaps, salves, lip balms, seeds (roasted & raw), food grade oil (bulk), hacky sacks, trade magazines, books and videos.

Dharma   Emporium
Rick Barbrick
3746 N. College, Indianapolis, IN 46205
USA
317-926-8255
Retailer
Products: Men, women and baby's clothing, diapers, handbags, wallets, accessories, hats, packs, outdoor gear, shoes, sandals, rope, twine, stationery, massage and body oils, salves, lip balms, seeds (raw), food grade oil (bulk), cake, cookies, hacky sacks, consumer periodicals, books, videos, jewelry.

The Dolphin Song
Linda S. Meisinger
102 S. Elm St., Gardner, KS
USA
913-856-7513   Fax: 913-856-7513
Retailer
Conscious environmental general store for social renewal.
Products: Men and women's clothing, handbags, wallets, accessories, hats, rope, twine, printing paper, hacky sacks, books and videos and more.

Dr. Brady's Hemp Seed Delights
Alan Brady
PO Box 43, Brookdale, CA 95007
USA
408-338
Manufacturer
Supplying hemp seed products made from raw hemp products.
Products: Salad oils and dressings, frozen deserts, candy, ice cream.

Earth Care
555 Leslie St., Ukiah, CA 95482-8507
USA
707-468-9214   Fax: 707-468-9486
Toll Free: 800-347-0070
Mail Order Catalog
Products: Men and women's clothing, hats, accessories, wash cloths, washmits, skin-care sets, salves, stationery, briefcases, portfolios.

Earth Goods USA, Inc.
David Edwards
2802 E. Madison #105, Seattle, WA 98112
USA
206-621-7757 Fax: 206-621-7785
Manufacturer
Natural clothing company using hemp dyed
with natural dyes.
Products: Men and women's clothing,
handbags, wallets, accessories.

Earth Pulp and Paper
John Stahl
PO Box 64, Leggett, CA, 95585
USA
707-925-6494 Fax: 707-925-6472
E-mail: tree@igc.apc.org
Manufacturer's Representative and
Distributor
Representative for pulping equipment and
distributor of pulp from hemp and other
alternative fibers. Will soon introduce a new
closed system pulping technology for hemp
and other non-wood fibers that uses no
chlorine, generates no toxic waste and has
reduced costs and higher yields by using
whole stalks.
Products: Pulping equipment, hemp pulp,
and other alternative fibers.

Earth Wish
Betty Yee
479 Harvard St., Brookline, MA, 02146
USA
617-566-0029   Fax: 617-566-0029
Retailer
Unusual, fun and affordable jewelry, clothing, bags, housewares, stationery made from recycled materials, cotton and hemp.
Products: Handbags, wallets, accessories, hats, printing paper, soaps, salves and lip balms.

Earthware's,   Inc.
Karin Yates
Carr Mill Mall, 101 E. Weaver St., Carrboro, NC 27510
USA
919-929-7844
Retailer
"Green" store selling planet friendly and conscious products.
Products: Men and women's clothing, handbags, wallets, accessories, hats, packs, outdoor gear, shoes, sandals, futons, rope, twine, printing paper, massage and body oils, salves, lip balms, seeds (raw).

Eastwinds Trading Company
Gary
PO Box 41, Obrien, OR 97534
USA
503-596-2331   Fax: 503-596-2331
Wholesaler, Importer
Hand-woven and crocheted hemp material
and accessories from Thailand.
Products: Men and women's clothing,
handbags, wallets, accessories, hats, packs,
outdoor gear, fabric, hacky sacks.

Eco Goods
Roger Berke
3555 Clares St. #X, Capitola, CA, 95010
USA
408-479-9293   Fax: 408-685-2919
Retailer
Complete "green" store selling all types of
earth friendly products.
Products: Men and women's clothing,
handbags, wallets, accessories, hats, packs,
outdoor gear, shoes, sandals, rope, twine,
printing paper, soaps, shampoos, massage and
body oils, salves, lip balms, cosmetics, hacky
sacks, books and videos.

Eco-Wise
Michele Primeaux
1714A South Congress, Austin, TX, 78704
USA
512-326-4474   Fax: 512-326-4496
Retailer
One-stop Eco-shop.
Products: Men, women, and children's
clothing, diapers, handbags, wallets,
accessories, hats, packs, outdoor gear, shoes,
sandals, fabric, rope, twine, printing paper,
soaps, shampoos, massage and body oils,
salves, lip balms, cosmetics, hacky sacks, art
supplies, paints, varnishes, trade magazines,
consumer periodicals, books and videos.

Ecological Wisdom
Lisa Wisdom
1705 N. 45th St., Seattle, WA, 98103
USA
206-548-1334   Fax: 206-548-1248
Retailer
Established to improve the environment by
offering environmental products including
hemp, organic cotton, energy and water
savings items and other products.
Products: Men and women's clothing,
handbags, wallets, accessories, hats, packs,
outdoor gear, fabric, rope, twine, printing
paper, soaps, salves, lip balms, seeds
(roasted), animal bedding, books and videos.

Ecolution
Eric Steenstra
PO Box 2279, Merrifield, VA, 22116
USA
703-207-9001   Fax: 703-560-1175
Importer, Distributor
Full line of hemp products including 100%
hemp jeans and jackets.
Products: Men and women's clothing,
handbags, wallets, accessories, hats, packs,
outdoor gear, fabric, rope, twine, canvas,
tarps, printing paper, salves, lip balms, seeds
(raw), food grade oil (bulk), hacky sacks.

Ecosource Paper, Inc.
Odette Kalman
111-1841 Oak Bay Ave., Victoria, BC V8R 1C4
Canada
604-595-4367   Fax: 604-370-1150
Toll-free: 800-665-6944
Importer
Tree-free hemp papers made from hemp and
other natural fibers.   Largest company in
North America with high volume for the
printing industry.
Products: Printing paper, specialty papers,
massage and body oils, seeds (raw).

The Emperor's Clothes
Holly Schroeder
1527 Defoe St., Missoula, MT, 59802
USA
406-728-3149
Retailer, Distributor, Manufacturer
Simply designed high quality clothing.
Products: Men and women's clothing, hats.

Emperor's Clothing Company
Martin Moravcik
133 Albert St., Winnipeg, Manitoba
Canada
204-947-2315 Fax: 204-956-5984
Toll-free: 800-665-HEMP
Retailer, Distributor, Importer, Manufacturer,
Sourcing Consultants.
Full line of hemp products.
Products: Men and women's clothing,
handbags, wallets, accessories, hats, shoes,
sandals, packs, outdoor gear, rope, twine,
printing paper, soaps, salves, lip balms,
granola, seeds (raw), hacky sacks.

EnvironGentle
Torrey
246 N. Hwy 101, Encinitas, CA, 92024
USA
619-753-7420
Toll-free: 800-6-NATURAL
Retailer
Earth-friendly gifts and supplies including a
large hemp product selection.

Evanescent Press
John Stahl
PO Box 64, Leggett, CA, 95585
USA
707-925-6494 Fax: 707-925-6472
E-mail: tree@igc.apc.org
Manufacturer
Make handmade paper from hemp (including local hemp) and other alternative fibers. Also, letterpress printing and hand bookbinding.
Products: Paper and related products.

Everything Earthly
Richard Scott
414 S. Mill Ave., #118, Tempe, AZ 85281
USA
602-968-0690 Fax: 602-968-0098
Retailer
Largest hemp store retailer in the Southwestern USA. Jewelry available at wholesale prices. Products also available by mail order.
Products: Men, women and baby's clothing, handbags, wallets, accessories, hats, packs, outdoor gear, shoes, sandals, wash cloths, fabric, rope, twine, stationery, soaps, massage and body oils, salves, lip balms, seeds (raw), food grade oil (bulk), cookies, candy, vitamins, supplements, consumer periodicals, books, videos, jewelry.

Exotic Gifts
J. Baker
PO Box 4665, Arcata, CA, 95521
USA
707-445-8981   Fax: 707-445-9214
Retailer
Products: Men and women's clothing,
handbags, wallets, accessories, hats, packs,
outdoor gear, sheets, pillows, futons, fabric,
rope, twine, printing paper, massage and body
oils, salves, lip balms, seeds (roasted and
raw), food grade oil (bulk), fiberboard,
insulation, caulking, putty, stucco, mortar,
cement blocks.

Exquisite Products Company
Ernest Orsi
10450 Wilshire Blvd., Los Angeles, CA
USA
310-470-9359   Fax: 310-475-9890
Importer
Importer of hemp products from Hungary and
China.
Products: Rope, twine, canvas, tarps, fabrics.

**First Hungarian Hemp Spinning
Company**
Toth Gyorgy
6724 Szeged, Londoni krt. 3
Hungary
36-62-313-830 Fax: 36-62-311-665
Manufacturer

120 year-old business of traditional binding items, fabric, canvas, insulation, oil and other hemp products.
Products: Handbags, wallets, accessories, towels, wash cloths, fabric, canvas, tarps, rope, twine, carpets, insulation, cold-pressed raw hempseed oil.

Forbidden   Fruits
Juan Torres
12837 Arroyo St., Sylvan, CA 91342
USA
800-258-1999   Fax: 818-837-1269
Wholesaler
Products: Soaps, shampoos, massage and body oils, salves, lip balms, seeds (roasted and raw), jewelry, pens.

Friendly   Stranger
Robin or Joy
226 Queen St. W. (upstairs), Toronto
Canada
416-591-1570   Fax: 416-591-1570
Retailer
Toronto's first and finest culture shop specializing in hempen goods and fabrics.
Products: Handbags, wallets, accessories, hats, fabric, rope, twine, rolling papers, hacky sacks, trade magazines, consumer periodicals, books and videos.

Greater Goods
Joan Kleban
515 High St., Eugene, OR, 97401
USA
541-485-4224   Fax: 541-485-8253
Retailer
A great selection of hemp goods including adult and infant clothing.
Products: Men, women and children's clothing, handbags, wallets, accessories, hats, packs, shoes, twine, writing paper, hacky sacks, trade magazines.

Green Machine
Philip Thomson
Hoogte Kadyk 53, 1018 BE Amsterdam
Holland
31-20-638-1096   Fax: 31-20-6382375
Manufacturer, Distributor, Sourcing Consultants.
Green Machine is a European-based manufacturer and distributor of ecological hemp products. Product range covers apparel and industrial textiles to paper, oils and foods.
Products: Men and women's clothing, handbags, wallets, accessories, hats, shoes, sandals, fabric, rope, twine, canvas, tarps, specialty papers, cardboard, packaging, massage and body oils, granola, seeds (raw), food grade oil (bulk), salad oils, dressings, candy, hacky sacks.

Greener   Alternatives
Bob Schwarz
914 Missions St. #A, Santa Cruz, CA, 95060
USA
408-423-0701   Fax: 408-423-0702
E-mail: ecodepot@greener.com
Retailer, Distributor
Hemp/Eco store and distributor of non-sterile
oil and other green products.
Products: Men and women's clothing,
handbags, wallets, accessories, hats, packs,
outdoor gear, fabric, rope, twine, printing
paper, technical papers, soaps, shampoos,
massage and body oils, salves, lip balms,
cosmetics, trade magazines, consumer
periodicals, books, videos.

Green   Underworld
Alethea Patton
6 Wharf Rd., Bolinas, CA 94924
USA
415-868-2531
Retailer
Store specializing in hemp and recycled
products.
Products: Men and women's clothing,
handbags, wallets, accessories, hats, packs,
outdoor gear, pillows, fabric, rope, twine,
printing paper, rolling papers, soaps, salves,
lip balms, consumer periodicals, books, videos,
jewelry.

Group W Bench
Raffael
1171 Chapel St., New Haven, CT 06511
USA
203-624-0683
Retailer
Pleasing the senses since 1968.
Products: Men, women and baby's clothing, diapers, handbags, wallets, accessories, hats, packs, outdoor gear, shoes, sandals, rope, twine, stationery, rolling papers, soaps, shampoos, massage and body oils, salves, lip balms, seeds (roasted), candy, hacky sacks, books, videos, jewelry.

Hayward Hempery
Robert Wilson
22544 Main St., Hayward, CA 94541
510-JET-WEED   Fax: 510-JET-WEED
USA
Retailer
Products: Men's clothing, handbags, wallets, accessories, hats, packs, outdoor gear, fabric, printing paper, soaps, shampoos, massage and body oils, salves, lip balms, cosmetics, salad oils, dressings, hacky sacks, art supplies, birdseed.

Head Trips Hat Company
Michele Woodward
372 Chrisman Ave., Ventura, CA, 93001
USA
805-643-3106
Wholesaler, Manufacturer
Items hand crocheted and crafted from hemp
twine. Catalog available.
Products: Women's clothing, hats, handbags,
rope, twine.

Hello Again
Sandie
14967 1st St. East, Madeira Beach, FL 33708
USA
813-824-7956
Manufacturer
Hand-dyed hemp yarn, hand-made. Custom
orders.
Products: Hats, wash cloths, belts, jewelry
(stock and customized), hacky sacks.

Hemcore, Ltd.
S. Carpenter
Station Rd., Felsted, Dunmow, Essex
England
44-0-1371-820066   Fax: 44-0-1371-820069
Retailer, Manufacturer, Distributor
The only company in the UK growing and
processing hemp fiber suitable for textiles
and paper. Core is available for bedding and
construction.

Products: Men and women's clothing, handbags, wallets, accessories, hats, shoes, sandals, fabric, rope, twine, carpets, animal feed and bedding, raw fibre and pulp.

Hemp BC
Marc Emery
324 W. Hastings, Vancouver, BC Z6B 1K6
Canada
604-681-4620 Toll Free: 800-330-HEMP
Retailer
Also available by mail order.
Products: Men and women's clothing, handbags, wallets, accessories, hats, packs, outdoor gear, shoes, sandals, fabric, stationery, rope, twine, printing paper, rolling papers, soaps, shampoos, massage and body oils, salves, lip balms, cosmetics, seeds (roasted and raw), food grade oil (bulk), cookies, candy, hacky sacks, consumer periodicals, books and videos.

The Hemp Club, Inc.
Larry Duprey
3418 A. Park Ave., Montreal, Quebec H2X 2H5
Canada
514-845-4993 Fax: 514-487-0651
Distributor, Manufacturer, Importer.
Products: Men and women's clothing, handbags, wallets, accessories, hats, packs, outdoor gear, shoes, sandals, fabric, rope, twine, printing paper, technical papers, soaps, massage and body oils, salves, lip balms, seeds.

The Hemp Connection
Marie Mills
782 Locust St., Garberville, CA, 95542
USA
707-923-4851
Retailer, Wholesaler, Manufacturer
Designs and fabricates hemp clothing for men and women.
Products: Men and women's clothing, pants, jackets, dresses, vests, shorts, skirts, and aprons.

Hemp Cooperation
Jeri Rose
PO Box 742, Redway, CA 95560
USA
707-923-5044
Consultant
Naturopathic Doctor and hemp activist. Also available as a speaker.
Products: Sourcing Consultant.

Hemp Educational Research Board
A. Das
PO Box 7137, Boulder, CO, 80306
USA
970-225-8356 Fax: 303-278-0560
Consultant, Research Foundation.
Scientific educational foundation which researches all aspects of the hemp plant, funds projects, solicits project proposals, and provides quality information for policy makers.

Products: *Hemp Fuels Digest: Food, Fiber, Fuels* and *Freedom for a Sustainable Future* and other related publications.

Hemp Essentials
Carol Miller
PO Box 151 Cazadero, CA 95421
USA
707-847-3642
Manufacturer
Natural body care products.
Products: Healing salves, lip balms, lotions, butter, massage oils, soaps, cosmetics, shampoos, herbal bath bags and dream pillows.

Hemp, Etc.
Shawn McMillian
37505 Eiland Blvd., Zephyr Hills, FL 33541
USA
813-782-6467 Fax: 813-782-6467
Retailer, Manufacturer
Also Available through mail order.
Products: Handbags, wallets, accessories, hats, packs, outdoor gear, cosmetics, beverage covers, sunscreens, sun blocks.

Hempfully Yours
Sarah Hutt
PO Box 1424, Forestville, CA 95436
USA
707-887-7741
Retailer
Unique high-fashion clothing and accessories made and sold directly to the user.
Products: Men, women and baby's clothing, handbags, wallets, accessories, hats, jewelry.

Hemp Head
Brandi Parfitt
4578 Queen St., Niagara Falls, ON L2E 2L6
Canada
905-371-1833
Retailer
Products also available by mail order.
Products: Men and women's clothing, handbags, wallets, accessories, hats, rope, twine, stationery, rolling papers, soaps, massage and body oils, salves, lip balms, seeds (raw), cookies, candy, consumer periodicals, books and videos.

Hemp Hemp Hooray
Lyn deMoss
PO Box 731, Occidental, CA, 95465
USA
707-874-2841
Manufacturer.
Products: Pet toys and bath accessories from hemp: bath mitts, wash cloths and soaps.

Hemp Hop
Jeff Gottsfeld
176 Madison Avenue, 4th floor., NY, NY 10016
USA
212-779-3265   Fax: 212-779-3255
Retailer, Distributor, Manufacturer.
Hemp related wear designed to raise consciousness and increase peace on the Earth.
Products: Men and women's clothing, hats and other miscellaneous apparel.

Hemp Hound
PO Box 1167, Occidental, CA, 95465
USA
707-874-1104   Fax: 707-824-0349
Manufacturer
100% hemp webbing and brass hardware are used in our dog collars and leashes.
Products: Dog collars and leashes.

Hemp is Hep
Somayah Kambui
826 W. 40th Pl., Los Angeles, CA 90037
USA
213-234-8701
Retailer, Distributor
Also available through mail order.
Products: Men and women's clothing, oils.

Hemp Magazine
Richard Tomcala
1304 W. Alabama, Houston, TX 77006
USA
713-521-1134   Fax: 713-528-HEMP
Publisher
A monthly publication on the many sides of the hemp issue.
Product: Publication: Hemp Magazine

Hemp On-Line
Alan Silverman
PO Box 14627, Santa Rosa, CA 95402
USA
707-579-8443
E-mail: alan@walstib.com
Consultant, Writer.
Assists hemp users and activists with productive use of the Internet and World Wide Web. Also writer for Hemp World Magazine and website.
Products: Information and connectivity.

Hemp Sacks
Kevin Johnson
690 Nature Lane, Arcata, CA 95521
USA
707-822-6972
Manufacturer
The original 100% hemp hacky sack.
Product: Hacky sacks.

Hempstead Company
Chris Boucher
2060 Placentia #B2, Costa Mesa, CA 92627
USA
Toll-free 800-284-HEMP   Fax: 714-650-5853
Manufacturer, Importer, Sourcing Consultants
First company in 40 years to grow industrial hemp in the United States.
Products: Men and women's clothing, handbags, wallets, accessories, hats, packs, outdoor gear, pillows, fabric, marine products, canvas, stationery, printing paper, postcards, soaps, shampoos, massage and body oils, salves, lip balms, seeds (raw), food grade oil (bulk), cookies, vitamins, supplements, hacky sacks, books, videos, jewelry, import brokerage.

Hempstead Company Store
Rob Boucher
601 Chartres St., New Orleans, LA 70130
USA
504-529-4367
Retailer.
Products: Men and women's clothing, handbags, wallets, accessories, hats, packs, outdoor gear, shoes, sandals, sheets, pillows, rope, twine, stationery, rolling papers, soaps, shampoos, massage and body oils, salves, lip balms, cosmetics, seeds (roasted), cookies, hemp honey, trade magazines, consumer periodicals, books, jewelry.

The Hemp Store
Robin Smith
1304 W. Alabama, Houston, TX, 77006
USA
713-523-3199   Fax: 713-528-HEMP
Distributor
Wholesale supermarket of hemp products for retailers interested in expanding their selection and variety.

Hemptech
John Roulac
PO Box 820, Ojai, CA, 93024
USA
805-646-HEMP Fax: 805-646-7404
Toll-free: 800-265-HEMP
E-mail: www.hemptech.com.
Publisher
The industrial hemp information network.
Products: Publishers of *Industrial Hemp* and *Practical Products--Paper to Fabric to Cosmetics, Bioresource Hemp Proceedings of the Symposium.*

Hemp Textiles International
David R. Gould
3200 30th St., Bellingham, WA, 98225
USA
360-650-1684 Fax: 360-650-0523
Toll-free: 800-778-4367
Manufacturer, Importer, Wholesaler.
Custom hemp and hemp/wool fabrics, hemp/organic cotton t-shirts and socks,

domestic spinning, weaving and knitting. Importer and wholesaler of Chinese hemp fabrics.
Products: Men and women's clothing, fabric.

Hemptown
Rose Phillips
232 N. LBJ, San Marcos, TX 78666
USA
512-396-0580   Fax: 512-396-0580
Retailer
Hemp product supplier for the Texas hill country.
Products: Men, women and children's clothing, handbags, wallets, accessories, hats, packs, outdoor gear, guitar straps, shoes, sandals, wash cloths, fabric, rope, twine, stationery, printing paper, rolling papers, soaps, massage and body oils, salves, lip balms, seeds (raw), food grade oil (bulk), cookie mixes, pancake mixes, hacky sacks, jewelry, consumer periodicals, books, videos.

Hemp   Traders
Lawrence Serbin
2130 Colby Ave., #1, Los Angeles, CA 90025
USA
310-914-9557   Fax: 310-478-2108
Importer, Exporter, Wholesaler.
Largest selection of hemp textiles in the world.
Products: Fabric, canvas and tarps.

Hemp Works, Inc.
David Marks
PO Box 648, Alpine, NJ, 07620
USA
201-784-5054
Toll-free: 800-715-6310
Retailer, Distributor.
Mail order distribution.
Products: Men and women's clothing, handbags, wallets, accessories, hats, packs, outdoor gear, shoes, sandals, futons, fabric, rope, twine, printing paper, soaps, massage and body oils, seeds (raw), salad oils, dressings, hacky sacks.

Hemp World Magazine
Mari Kane
PO Box 315, Sebastopol, CA 95473
USA
707-887-7508   Fax: 707-887-7639
E-mail: hemplady@crl.com
Web: http://hempworld.com
Publisher.
An international hemp journal
Products: Trade magazines, books and videos.

Hempy's
Albert Lewis
600 Front St. #315, San Diego, CA 92101
USA
619-233-HEMP   Fax: 619-233-4367
Manufacturer, Wholesaler.
Quality hemp products made from the most
environmentally sound materials possible.
Products: Surfboard bags, full and mini-
sized backpacks, hats, men and women's
clothing, handbags and accessories.

Herbal Products and Development
Paul Gaylon
PO Box 1084, Aptos, CA, 95001
USA
408-688-8706   Fax: 408-688-8711
Distributor, Importer.
Extra virgin cold pressed hemp seed oil from
the Andes. Packaged products and bulk
drums available. Also, blended food oils for
food and cosmetic uses.
Products: Food grade oil (bulk), industrial
lubricants, refined oil, supplements, salad oils.

Hip Hemp
Krissi Handwerger
PO Box 4839, Pittsburgh, PA, 15206
USA
412-734-5538
Retailer, Manufacturer, Distributor.
Quantity producers and distributors of
affordable, quality hemp products since 1991.

Products: Jewelry, fanny packs, and handbags, hats, hacky sacks.

Home Grown Hats
Jane Dawson
PO Box 1083, Redway, CA, 95560
USA
707-923-5273 Fax: 707-923-5273
Manufacturer, Wholesaler.
Largest selection of hat styles and colors.
Products: Hats.

Humboldt Industrial Hemp Supply Co.
J. Baker
PO Box 4665, Arcata, CA, 95521
USA
707-445-8981 Fax: 707-445-9214
Distributor, Manufacturer, Sourcing
Consultants, Import Brokers, Contract Fabric
and Garment Dyeing.
Supplier of hemp fabric, hemp fiber, building
supplies.
Products: Men and women's clothing,
handbags, wallets, accessories, hats, packs,
outdoor gear, sheets, pillows, futons, fabric,
rope, twine, printing paper, massage and body
oils, salves, lip balms, seeds (roasted and
raw), food grade oil (bulk), fiberboard,
insulation, caulking, putty, stucco, mortar,
cement blocks.

Hungry Bear Hemp Foods
Todd Dalotto
PO Box 12175, Eugene, OR, 97440
USA
541-345-5216   Fax: 541-302-1488
Manufacturer.
Producers of Seedy Sweeties (vegan hempseed treats) and other fine foods. Distributor of hempseed oil pressed from non-sterilized seed.
Products: Candy, treats, seeds (raw), food grade oil (bulk).

Institute For Hemp
PO Box 65130, St. Paul, MN, 55165
USA
612-222-2628   Fax: 612-222-2628
Association, Consultants.
Non-profit organization for the promotion of industrial hemp.
Products: Trade magazines, books and videos.

International Hemp Association
David Watson
Postbus 75007, 1070AA, Amsterdam
Netherlands
131-0-20-618-8758   Fax: 131-0-20-618-8758
Association, Publisher.
Disseminators of hemp publications and educational materials towards the advancement of the use of hemp worldwide.
Products: Peer-reviewed scientific journal.

Labyrinth Phassions and Costumes
Jennifer Jensen
463 Haight St., San Francisco, CA, 94117
USA
415-552-3082 Fax: 415-552-3083
Manufacturer.
Hauntingly classical to playfully psychedelic hemp phassions for men and women.
Products: Men and women's clothing, hats.

Longevity Book Arts
Jonathan Root, Niya Nolting
PO Box 1392, Eugene, OR, 97440
USA
Toll-free: 800-334-1952
Manufacturer, Distributor.
Traditional book bindery which offers the first line of hemp sketch books and journals including Soft Sativa, the only 100% hemp book.
Products: Sketchbooks, journals, custom-made hemp books.

Lost Harvest
Eric Pierce
PO Box 615, Rye, NH, 03870
USA
603-431-5966 Fax: 603-431-1489
Manufacturer.
Offers a variety of colorful hemp products from hats to hacky sacks all made in the U.S.

Products: Handbags, wallets, accessories, hats, packs and outdoor gear, shoes, sandals, hacky sacks, chalk bags, jewelry.

Makop Linen Mill
Swietiana Gorka
Malbork 82-200, ul. Daleka 115
Poland
48-55-33-21 or 29   Fax: 48-55-30-77
Manufacturer
Linen and hemp fibers, wet and dry spun linen and hemp worsted and carded yarns. Can also do boiling and bleaching.
Products: Linen and hemp yarns.

Magna International, Inc.
Yola Lawska
6704 S. 239th Pl, #B104, Kent, WA, 98032
USA
206-813-8429 Fax: 206-813-8735
Importer, Wholesaler.
Hemp twine and fabric imported from Poland and Turkey
Products: Rope, twine, tarps, canvas, fabric.

Mama Indica's Hemp Seed Treats
Ron Shaul
Box 591, Tofino, BC V0R 2Z0
Canada
604-725-4288 Fax: 604-725-4288
Manufacturer
Makers of fine hemp products since 1992.

Products: Seeds (roasted and raw), hemp seed treats.

Manashtash, Inc.
Robert E. Jungmann
1915 21st Ave. South, Seattle, WA, 98144
USA
Manufacturer, Distributor.
Textiles producers and distributors of environmentally friendly products for the future of the Earth.
Products: Men, women and children's clothing, handbags, wallets, accessories, packs, outdoor gear, sheets, pillows, towels, wash cloths.

Mary Jane's Hemp Harvest
Tonya, Dene, Bob
4514 Central SE, Albuquerque, NM, 87108
USA
505-254-0000
Retailer.
Natural hemp fiber apparel and accessories. Largest hemp store in New Mexico.
Products: Men, women and children's clothing, handbags, wallets, accessories, hats, packs, outdoor gear, shoes, sandals, sheets, pillows, fabric, printing paper, rope, twine, soaps, shampoos, body oils, salves, lip balms, cosmetics, seeds (roasted), food grade oil (bulk), vitamins, supplements, hacky sacks, trade magazines, consumer periodicals, books, videos, candles, dream catchers, jewelry.

Mary Jane's Hemp Seed Snacks
Annie Riecker
NZ Vonbinswal 66 #70, 1012 SC Amsterdam
Netherlands
31-20-673-9541   Fax: 31-20-673-5910
Manufacturer.
Toasted hemp seeds, chocolate hempseed
clusters made with live seed.
Products: Seeds (roasted), candy, cookies.

MAterials
Jay and Kerri Fickes
PO Box 741, San Luis Obispo, CA, 93406
USA
805-595-2056 Fax: 805-595-2058
Toll-free: 800-OK-BUY-MA
Retailer, Distributor, Manufacturer.
Large selection of men, women and children's
clothing and accessories. Fashionable
designer wear.
Products: Men, women and children's
clothing, handbags, wallets, accessories, hats,
packs, outdoor gear.

The Message Company
James Berry
4 Camino Azul, Santa Fe, NM 87505
USA
505-474-0998   Fax: 505-471-2584
Publisher
Publisher of the *International Directory of Hemp Products and Suppliers* and other books and videos on spirituality in business, making money, freedom, common law, legal self-help, new energy, new science, lost technologies and emergency preparedness.
Products: Books and Videos.

The Merry Hempsters
Jerry Shapiro
PO Box 1301, Eugene, OR 97440
USA
541-345-9317   Fax: 541-345-0910
Manufacturer
An eco-conscious manufacturer.
Products: Lip balms, cosmetics, seed cake, seed meal, food grade oil (bulk), industrial grade oil (bulk).

Mindful Products
Max Salkin
20095 First Street West, Sonoma, CA, 95476
USA
707-939-9161   Fax: 707-939-9161
Manufacturer.
Produce fine hemp clothing.

Products: Men and women's clothing, tennis shoes from 100% hemp canvas.

Montana Hemp Traders
Jackie Beyer
1455 Ft. Macleod Trail, Eureka, MT, 59917
USA
406-889-3091
Retailer, Distributor.
Products: Men and women's clothing, handbags, wallets, accessories, hats, packs, outdoor gear, fabric, rope, twine, printing paper, soaps, massage and body oils, salves, lip balms, salad oil, dressings, hacky sacks, hemp seed treats, books, videos.

Mystic Eye
Laura Petritz
30 N. Lexington Ave., Asheville, NC, 28801
USA
704-251-1773   Fax: 704-254-7655
Toll-free: 800-345-3999
Retailer, Contract Dyeing.
Retail store and art gallery specializing in original artwork, hemp products and garment and fabric dyeing.
Products: Men and women's clothing, handbags, wallets, accessories, shoes, sandals, rope, twine, massage and body oils, salves, lip balms, contract fabric and garment dyeing.

Nagylaki Kenderfono Gyar Rt.
Mr. Kiss
6933 Nagylak Gyar Str. 1.
Hungary
36-62-411557   Fax: 36-62-411394
Manufacturer
Planter and grower of hemp. Producer of fibers, packaging and hemp baler twine, chipboard.
Products: Fabric, twine, packaging, cardboard, fiberboard.

Natural Choice
1365 Rufina Circle, Santa Fe, NM 87505
USA
505-438-3448   Fax: 505-438-0199
Toll Free: 800-621-2591
Retailer, Mail Order Catalog.
Products: Backpacks, fanny packs, briefcases, duffel bags.

Natural Hemphasis
David Marcus
833 Manning Ave., Toronto, Ontario M6G 2W9
Canada
416-535-3497
E-mail: nathemp@interlog.com
Distributor, Manufacturer
Products: Handbags, wallets, accessories, rolling papers, lamp shades, frisbees.

Natural Rags and Hempery
Sue Lukasha
212 Main St., Nevada City, CA 95959
USA
916-265-5545   Fax: 916-274-7357
Retailer.
Products: Men and women's clothing,
handbags, wallets, accessories, hats, packs,
outdoor gear, shoes, sandals, fabric, rope,
twine, printing paper, rolling papers, soaps,
massage and body oils, salves, lip balms,
seeds (roasted and raw), food grade oil (bulk),
hacky sacks, birdseed, trade magazines, books
and videos.

Naturetex   International   BV
Michael Rich
Van Diemenstraat 192, 1013 CP Amsterdam
Netherlands
31-0-20-420-3040   Fax: 31-0-20-420-3545
Manufacturer, Importer.
Manufactures pure hemp and hemp blend
textiles in China and imports them into
Europe and North America.
Products: Fabric.

Naturetex   NYC
Lisa Rich
PO Box 536, Westbrook, CT, 06498
USA
203-399-5355   Fax: 203-399-5313
Importer.

Sales representatives of all hemp textile products for Naturetex International BV.
Products: Fabric.

The New Age Emporium
Chuck Whitehouse
22 Washington St., Camden, ME, 04843
USA
207-236-8628
Retailer.
Environmentally sound products and tools for healthy living.
Products: Men, women and children's clothing, sheets, pillows, towels, wash cloths, specialty papers, soaps, shampoos, massage and body oils, salves, lip balms, cosmetics, salad oils, dressings, vitamins, supplements, art supplies, books and videos.

Of The Earth
Richard Ziff
916 W. Broadway #749, Vancouver, BC V5Z 1K7
Canada
604-878-1268 Fax: 604-878-1268
Manufacturer
100% hemp clothing, bedding and accessories with natural dyes.
Products: Men, women and children's clothing, handbags, wallets, accessories, hats, packs and outdoor gear, sheets and pillows, futons, fabric.

The Ohio Hempery, Inc.
David Smigelski
7002 SR 329, Guysville, OH, 45735
USA
614-662-4367   Fax: 614-662-6446
Toll-free: 800-BUY-HEMP
Retailer, Distributor, Manufacturer, Importer,
Sourcing Consultants.
A complete supplier of hemp products.
Products: Men, women, children's clothing,
diapers, handbags, wallets, accessories, hats,
packs, outdoor gear, shoes, sandals, fabric,
webbing, rope, twine, canvas, tarps, printing
paper, specialty papers, soaps, shampoos,
body oils, salves, lip balms,. seeds (raw), food
grade oil (bulk), vitamins, supplements, hacky
sacks, printing inks, paints, varnishes, feed
and animal bedding, books, videos, raw
materials (yarn, sliver, tow, hurds, stalk,
bark). Contract fabric and garment dyeing.

Ohotto Hemp
Jeff Ohotto
22 99th Ln., Coon Rapids, MN, 55448
USA
612-786-4220
Manufacturer, Distributor.
Products: Men, women and children's
clothing, handbags, wallets, accessories, hats,
packs, pillows.

Ohotto Hemp Retail Store

Jeff Ohotto
420 E. Main St. #17, Anoka, MN, 55303
USA
612-576-0046   Fax: 612-576-6614
Retailer.
Products: Men, women and children's clothing, handbags, wallets, accessories, hats, packs, pillows, twine, soaps, salves and lip balms, hacky sacks, books and videos.

Original   Sources

A. Das
PO Box 7137, Boulder, CO, 80306
USA
970-225-8356   Fax: 303-278-0560
Manufacturer, Sourcing Consultants.
Food, fiber and fuels for a sustainable planet. Developing infrastructure in preparation for domestic cultivation. Also consulting and technical support in many areas.
Products: Paper pulping machinery, massage and body oils, granola, bulk hemp seeds (roasted and raw), seed cleaning equipment, cookies, food grade oil (bulk), salad oils, dressings, hemp flour, vitamins, supplements, art supplies, lubricants, fuel, solvents and coatings, hemp fiber reinforced plastics, lubricants, biomass fuels, gasifiers, birdseed, feed, fertilizer, books and videos.

Pan World Traders
Dylan
PO Box 697, Santa Cruz, CA, 95061
USA
408-479-4803   Fax: 408-476-5965
Manufacturer, Importer.
Direct weaver and finished goods
manufacturer in Romania. Custom orders.
Products: Men, women and children's
clothing, diapers, handbags, wallets,
accessories, hats, packs, outdoor gear, kitchen
ware, placemats, hot pads, pot holders,
aprons, fabric, rope, twine, tarps, canvas,
carpets, hacky sacks.

Perceptions
Gini Kramer-Goldman
10734 Jefferson Blvd. #502, Culver City, CA,
90230
USA
310-313-5185   Fax: 310-313-5198
Toll-free: 800-276-4448
Publisher.
Magazine covering interesting information
about the hemp industry, government, health
and metaphysics through alternative news.
Products: Consumer periodical.

Pickering   International
Dawn Pickering
888 Post Street, San Francisco, CA, 94109
USA
415-474-2288   Fax: 415-474-1617

Distributor, Importer, Sourcing Consultants, Import Brokerage.
Imports and distributes hemp fabrics and consults on sourcing hemp products.
Products: Men and women's clothing, shoes, fabric.

PlanetWise Products
Skip Broadhead
6824 S. 19th St. #126, Tacoma, WA 98466 USA
206-770-7296   Fax: 206-770-7396
Retailer.
Two retail stores in Federal Way and Puyalluk, Washington.   Full-line environmental products retailer. Carry hemp clothing and accessories.
Products: Men, women and children's clothing, handbags, wallets, accessories, hats, packs, outdoor gear, shoes, sandals, sheets, pillows, towels, wash cloths, soaps, shampoos, massage and body oils, salves, lip balms, hacky sacks, birdseed, fertilizer.

Quick Distribution Company
Dave Lewis
1626 East 22nd Street, Oakland, CA, 94606 USA
510-436-6291   Fax: 510-HEMP291
Distributor.
Products: Handbags, wallets, accessories, hats, salves, lip balms, hacky sacks, patches, rubber stamps, books and videos.

Real Goods
555 Leslie Street, Ukiah, CA 95482-5507
USA
707-468-9214   Fax: 707-468-9486
Toll Free: 800-762-7325
Mail Order Catalog.
Products: Paper, stationery, bags, shoes.

Rising Star Futon
Bill Kurtz
35 NW Bond St., Bend, OR, 97701
USA
503-382-4221   Fax: 503-383-5925
Toll-free: 800-828-6711
Retailer, Manufacturer.
Custom cushions and futons (filled with
wellspring polyfiber from recycled soda
bottles) and covered with organic hemp twill
canvas.
Products: Futons and covers, sheets, pillows
and cushions.

Sexton Belt Company
Andrew Sexton
506 Lee Place, Frederick, MD 21702
USA
301-663-9497
Retailer
Maker of hand-made hemp twine belts.
Products: Belts.

Shakedown Street
Bob Lamar
276 King St. West, Kitchner, Ontario N2G 1B7
Canada
519-570-0440   Fax: 519-570-0440
Retailer.
Products: Men and women's clothing,
handbags, wallets, accessories, hats, packs,
outdoor gear, rope, twine, stationery, rolling
papers, soaps, massage and body oils, salves,
lip balms, seeds (raw and roasted), hacky
sacks, consumer periodicals, books.

Sharon's Finest
Russ Postel
PO Box 5020, Santa Rosa, CA 95402
USA
707-576-7050   Fax: 707-545-7116
Manufacturer
Developing new and healthy food alternatives
containing nutritious and delicious hemp
seeds and oil.
Products: Hemp cheese, hemp burgers,
frozen deserts.

Simply Better
John Quinney
90 Church Street, Burlington, VT, 05401
USA
802-658-7770   Fax: 802-658-2360
Retailer.
Environmental store offering more than 2000
products.

Products: Men and women's clothing, handbags, wallets, accessories, packs, outdoor gear, rope and twine.

Solar Age Press
Jack Frazier
PO Box 610, Peterstown, WV, 24963
USA
No Phone
Publisher.
Solar Age Press publishes pro-hemp literature.
Products: *Hemp Paper Reconsidered, The Great American Hemp Industry, The Solar Age Hemp Paper Report* (newsletter).

Sow Much Hemp
Bruce Mullican
245 W. 13th Ave., Eugene, OR 97401
USA
503-344-7200
E-Mail: hire@efn.org
Web: http://www.efn.org/~bgm/sowmuch
    hemp.html
Retailer, Distributor.
Also available through mail order catalog.
Products: Men, women and children's clothing, handbags, wallets, accessories, hats, packs, outdoor gear, shoes, sandals, pillows, bath mitts, wash cloths, fabric, rope, twine, stationery, printing paper, rolling papers, soaps, shampoos, body oils, salves, lip balms, cosmetics, seeds (raw and roasted), food grade

oil (bulk), hacky sacks, trade magazines, consumer periodicals, books, videos, jewelry.

Still Eagle Planetary
Nick Sminow
557 Ward St., Nelson BC V1L 1T1
Canada
604-352-3844   Fax: 604-352-9288
Retailer, Distributor.
One stop hemp products distributors.   First hemp store in British Columbia.
Products: Men, women and children's clothing, handbags, wallets, accessories, hats, packs, outdoor gear, shoes, socks, fabric, rope, twine, stationery, printing paper, rolling papers, soaps, shampoos, massage and body oils, herbal bath bags, salves, lip balms, seeds (roasted and raw), food grade oil (bulk), candy, vitamins and supplements, hacky sacks, consumer periodicals, books, videos, jewelry, hanging shelves.

Stradom S.A.
Mr. Warkiewicz, Mr. Gancarz
21 First May St., 42-200, Czestochowa
Poland
48-34-242-777 Fax: 48-34-651-495
Manufacturer.
Spin and weave hemp, flax, jute, cotton and polypropylene yarn into flat and tubular cloth.
Products: Combed hemp natural fibres, dry-woven hemp worsted yarn, technical flax and

hemp carded yarn, twisted hemp yarn, hemp binder string, hemp cordage, various fabrics, canvas, nets, hemp sailcloth, raw materials.

Sue's Amazing Lip Stuff
Sue Kastensen
PO Box 64, Westby, WI, 54667
USA
608-634-2988   Fax: 608-634-2988
Manufacturer.
Native American recipe used for all natural skin care products.
Products: Salves and lip balms.

Sunsports
Joseph N. Maxner
PO Box 180, Stamford, CT, 06904
USA
203-324-2191   Fax: 203-324-6651
Manufacturer, Distributor.
Manufacturing and distributing products made with 100% hemp. All products made in the United States.
Products: Handbags, wallets, accessories, hats, packs, outdoor gear, shoes, sandals.

Sweetlight Books
Guy Mount
16625 Heitman Rd., Cottonwood, CA, 96022
USA
916-529-5392
Publisher, Book Distributor.

Products: *Green Gold; Hemp Today; Hemp: Lifeline to the Future; Excerpt From the Indian Hemp Drugs Report* and others.

SWIHTCO, The Swiss Hemp Trading Co.
Shirin Patterson
Ch-3205 Mauss, 1000 Lausanne
Switzerland
41-31-751-30-05
Manufacturer.
Cultivation and commercialization of hemp primary materials.
Products: Seeds (raw), flowers, leaves, stalks.

Texas Hemp Company
Richard Tomcala
1304 W. Alabama, Houston, TX 77006
USA
713-521-1134 Fax: 713-528-HEMP
Toll-free: 800-506-HEMP
Retailer.
Oldest hemp store in America.

The Third Stone Hemp Products
Jay Sullivan
703 W. Lake St., Minneapolis, MN 55408
USA
612-825-6120 Fax: 612-825-7040
Retailer.
Also produce a three-day hemp festival called The Festival for Project Earth.

Products: Men, women and children's clothing, handbags, wallets, accessories, hats, packs, outdoor gear, shoes, sandals, pillows, wash cloths, fabric, rope, twine, stationery, printing paper, rolling papers, soaps, shampoos, massage and body oils, salves, lip balms, seeds (raw and roasted), food grade oil (bulk), salad oils, dressings, hacky sacks, trade magazines, consumer periodicals, books, videos, jewelry.

Tomorrow's World
Richard
1564 Laskin Rd., Virginia Beach, VA, 23451
USA
804-437-8911   Fax: 804-437-8051
Retail (mail order).
Products: Hemp clothing, shoes, chemically-free bedding and clothing.

Tomtex S.A.
Joanna Warczynska
97-200 Tomaslow Mazowiecki, UL,
Wkokiennicza 12/1E
Poland
48-45-23-3257 Fax: 48-45-23-3271
Manufacturer.
Products: Fabric.

Treefree Eco-Paper
David Potter
121 SW Salmon #1100, Portland, OR 97204
USA
Toll-free: 800-775-0225
Distributor, Manufacturer.
Manufacturer and distributor of non-wood papers..
Products: Stationery, printing paper, bible
and specialty papers, newsprint, card stock,
journals, notebooks, envelopes, sourcing
consultants, import brokerage of non-wood
pulp.

True North Hemp Company, Ltd.
Troy or Kevin Stewart
#103, 10324-82nd Ave., Edmonton, Alberta
T6E 1Z8
Canada
403-437-HEMP Fax: 403-437-4367
Retailer.
Advocate, lobbyist, and retailer of the
Canadian hemp movement.
Products: Men and women's clothing,
handbags, wallets, accessories, hats, packs,
outdoor gear, shoes, sandals, sheets, pillows,
fabric, rope, twine, canvas, tarps, printing
paper, technical papers, rolling papers, soaps,
massage and body oils, salves, lip balms,
granola, seeds (raw and roasted), salad oils,
dressings, hacky sacks, art supplies, trade
magazines, consumer periodicals, books and
videos.

Two Star Dog
Steven Boutrous
1370 10th St., Berkeley, CA, 94710
USA
510-525-1100   Fax: 510-525-8602
Manufacturer, Importer, Wholesaler.
Products: Men and women's clothing, hats,
packs and outdoor gear.

Ukrimpex
Mr. S. Sokolenko
22 Vorovsky Str. Kiev 252054
Ukraine
380-44-216-2174   Fax: 380-44-216-2996
Distributor, Importer, Sourcing Consultants,
Import Brokerage,.
First export-import company in Ukraine
dealing with hemp.
Products: Textiles, paper, food, building
materials, feed, supplier services.

U.S. Hemp
Kathy Trout
461 Apache Trail #130, Apache Junction, AZ
85220
USA
602-983-7065   Fax: 602-983-9785
E-mail: ushemp@ix.netcom.com
Retailer, Distributor, Manufacturer.
Handmade hemp products manufactured in
the United States.
Products: Men, women and children's
clothing, handbags, wallets, accessories, hats,

packs and outdoor gear, shoes, rope, twine, printing paper, stationery, soaps, shampoos, salves, lip balms, seeds (raw and roasted), food grade oil (bulk), cookies, jewelry.

U.S. Textiles
Eugene Sawicki
404 W. Pico Blvd., Los Angeles, CA 90015
USA
213-742-0840 Fax: 213-742-0016
Manufacturer, Importer.
Fabric imported from Poland, Russia, Kajakistan and the Ukraine. Manufacturers of hats, packs, t-shirts and pants.
Products: Men's clothing, hats, packs, outdoor gear, fabric.

Vermont Hemp Company
Larry Phillips
PO Box 5233, Burlington, VT 05402
USA
802-865-2646
Distributor, Wholesaler.
Wholesaler and supplier of bulk hemp food ingredients.
Products: Seeds (raw and roasted), food grade oil (bulk), hemp flour, hemp baking flour.

Wise Up Reaction Wear
Chris Ball
1511 W. Wetmore Rd., Tucson, AZ 85705
USA
520-293-8005   FAX: 520-293-2237
Toll-Free: 800-296-HEMP
Manufacturer
Hemp and related products for a free and sustainable future.
Products: Handbags, wallets, accessories, patches, salves, lip balms, jewelry.

Index by Country

Australia
Australian Hemp Products

Belgium
Alma Rosa N.V. Belgium

Canada
Chanvre en Ville
Cotton and Willow
Ecosource Paper, Inc.
Emperor's Clothing Company
Friendly Stranger
Hemp BC
The Hemp Club, Inc.
Hemp Head
Herban Home
Mama Indica's
Natural Hemphasis
Of The Earth
Shakedown Street
Still Eagle Planetary
True North Hemp Company, Ltd.

England
British Hemp Stores
Hemcore, Ltd.

Hungary

CEL BT

First Hungarian Hemp Spinning
Company

Nagylaki Kenderfono Gyar Rt.

Netherlands

Green Machine

International Hemp Association

Mary Jane's Hemp Seed Snacks

Naturetex International BV

Poland

Makop Linen Mill

Stradom S.A.

Tomtex S.A.

Switzerland

SWIHTCO, The Swiss Hemp Trading Co.

Ukraine

Ukrimpex

USA by State

Alaska
Alaska Green Goods

Arizona
Crucial Creations
Cultural Repercussions
Everything Earthly
Hemp Sprouts
U.S. Hemp
Wise Up Reaction Wear

California
2000 BC
AH Hemp
All Points East
Artha
Business Alliance for Commerce in
    Hemp
Cannabest
Cannabis Clothes
Canvas Hemp Company (CHC)
Community Market
Creative Expressions
Dr. Brady's hemp Seed Delights
Earth Care
Earth Pulp and Paper
Eco Goods
EnvironGentle
Evanescent Press
Exotic Gifts

Exquisite Products Company
Forbidden Fruits
Greener Alternatives
Green Monkey Creations
Green Underworld
Hayward Hempery
Head Trips Hat Company
The Hemp Connection
Hemp Cooperation
Hemp Essentials
Hempfully Yours
Hemp Hemp Hooray
Hemp Heritage
Hemp Hound
Hempire
Hemp is Hep
Hemp On-Line
Hemp Sacks
Hemp Shak
Hempstead Company
Hemptech
Hemp Traders
Hemp World Magazine
Hempy's
Herbal Products and Development
Home Grown Hats
Humboldt Industrial Hemp Supply Co.
Labyrinth Phassions and Costumes
MAterials
Mindful Products
Natural Rags and Hempery
Pan World Traders
Perceptions

Pickering International
Quick Distribution Company
Real Goods
Sharon's Finest
Sweetlight Books
Two Star Dog
U.S. Textiles

Colorado
The Boulder Hemp Company (AKA One Brown Mouse)
Hemp Educational Research Board
Original Sources

Connecticut
Group W Bench
Naturetex NYC
Sunsports

Florida
Hello Again
Hemp, Etc.

Georgia
AMPT

Hawaii
Grassroots of Hawaii

Illinois
Harvest House

Indiana
Dharma Emporium

Kansas
The Dolphin Song

Louisiana
Hempstead Company Store

Maine
The New Age Emporium

Maryland
Sexton Belt Company

Massachusetts
Earth Wish

Minnesota
Institute For Hemp
Ohotto Hemp
Ohotto Hemp Retail Store
The Third Stone Hemp Products

Montana
The Emperor's Clothes
Montana Hemp Traders

New Hampshire
Lost Harvest

New Jersey
Danforth International
Hemp Works, Inc.

New Mexico
Cannabiz Company
Mary Jane's Hemp Harvest
The Message Company
Natural Choice

New York
Aradia's Kitchen
Hemp Hop

North Carolina
Earthware's, Inc.
Mystic Eye

Ohio
The Ohio Hempery, Inc.

Oregon
AdventureSmiths
ASA Aware Products
Eastwinds Trading Company

Greater Goods
Hungry Bear Hemp Foods
Longevity Book Arts
Merry Hempsters
Rising Star Futon
Sow Much Hemp
Treefree Eco-Paper

Pennsylvania
Hip Hemp

Texas
Eco-Wise
Hemp Magazine
The Hemp Store
Hemptown
Texas Hemp Company

Vermont
Artisan Weavers
Green Horizons
Simply Better
Vermont Hemp Company

Virginia
Ecolution
Tomorrow's World

Washington
All Around the World Hemp
American Hemp Mercantile
Crop Circle Clothing

Deep See
Earth Goods USA, Inc.
Ecological Wisdom
Hemp Textiles International
Magna International, Inc.
Manashtash, Inc.
PlanetWise Products

West Virginia
Solar Age Press

Wisconsin
Sue's Amazing Lip Stuff

Index By Category

Retailers

Green Underworld
Group W Bench
Hayward Hempery
Hemcore, Ltd.
Hemp BC
The Hemp Connection
Hemp, Etc.
Hempfully Yours
Hemp Head
Hemp Hop
Hemp is Hep
Hemp Shak
Hemp Sprouts
Hempstead Company Store
Hemptown
Hemp Works, Inc.
Hip Hemp
Mary Jane's Hemp Harvest
MAterials
Montana Hemp Traders
Mystic Eye
Natural Choice
Natural Rags and Hempery
The New Age Emporium
The Ohio Hempery
Ohotto Hemp Retail Store
Planet Wise Products
Real Goods
Rising Star Futon
Sexton Belt Company
Shakedown Street
Simply Better
Sow Much Hemp

Still Eagle Planetary
Texas Hemp Company
The Third Stone Hemp Products
Tomorrow's World
True North Hemp Company, Ltd.
U.S. Hemp

Distributors/Wholesalers

AH Hemp
All Around the World Hemp
All Points East
Alma Rosa N.V. Belgium
American Hemp Mercantile
AMPT
Aradia's Kitchen
Australian Hemp Products
The Boulder Hemp Company
British Hemp Stores
Cannabest
Cannabiz Company
Crop Circle Clothing
Deep See
Earth Pulp and Paper
Eastwinds Trading Company
Ecolution
The Emperor's Clothes
Emperor's Clothing Company
Forbidden Fruits
Grassroots of Hawaii
Green Machine
Greener Alternatives
Harvest House

107

Head Trips Hat Company
Hemcore, Ltd.
The Hemp Club, Inc.
The Hemp Connection
Hemp Hop
Hemp is Hep
Hemp Sprouts
The Hemp Store
Hemp Textiles International
Hemp Traders
Hemp Works, Inc.
Hempy's
Herbal Products and Development
Herban Home
Hip Hemp
Home Grown Hats
Humboldt Industrial Hemp Supply Co.
Longevity Book Arts
Magna International, Inc.
Manashtash, Inc.
MAterials
Montana Hemp Traders
Natural Hemphasis
The Ohio Hempery
Ohotto Hemp
Pickering International
Quick Distribution Company
Sow Much Hemp
Still Eagle Planetary
Sunsports
Treefree Eco-Paper
Two Star Dog
Ukrimpex

U.S. Hemp
Vermont Hemp

Importers

American Hemp Mercantile
Australian Hemp Products
British Hemp Stores
Cannabiz Company
Cotton and Willow
Crop Circle Clothing
Danforth International
Eastwinds Trading Company
Ecolution
Ecosource Paper, Inc.
Emperor's Clothing Company
Exquisite Products Company
The Hemp Club, Inc.
Hempstead Company
Hemp Textiles International
Hemp Traders
Herbal Products and Development
Humboldt Industrial Hemp Supply Co.
Magna International, Inc.
Naturetex International BV
Naturetex NYC
The Ohio Hempery
Pan World Traders
Pickering International
Two Star Dog
Ukrimpex
U.S. Textiles

Manufacturers

AdventureSmiths
AH Hemp
All Around the World Hemp
American Hemp Mercantile
AMPT
Aradia's Kitchen
Artha
Artisan Weavers
ASA Aware Products
Australian Hemp Products
The Boulder Hemp Company
British Hemp Stores
Canvas Hemp Company
CEL BT
Crop Circle Clothing
Crucial Creations
Cultural Repercussions
Dr. Brady's Hemp Seed Delights
Earth Goods USA, Inc.
The Emperor's Clothes
Emperor's Clothing Company
Evanescent Press
First Hungarian Hemp Spinning
Grassroots of Hawaii
Green Horizons
Green Machine
Green Monkey
Harvest House
Head Trips Hat Company
Hello Again
Hemcore, Ltd.

The Hemp Club, Inc.
The Hemp Connection
Hemp Essential
Hemp, Etc.
Hemp Hemp Hooray
Hemp Heritage
Hemp Hop
Hemp Hound
Hempire
Hemp Sacks
Hempstead Company
Hemp Textiles International
Hempy's
Herban Home
Hip Hemp
Home Grown Hats
Humboldt Industrial Hemp Supply Co.
Hungry Bear Hemp Foods
Labyrinth Phassions and Costumes
Longevity Book Arts
Lost Harvest
Makop Linen Mill
Mama Indica's Hemp Seed Treats
Manashtash, Inc.
Mary Jane's Hemp Seed Snacks
MAterials
Merry Hempsters
Mindful Products
Nagylaki Kenderfono Gyar Rt.
Natural Hemphasis
Naturetex International BV
Of The Earth
The Ohio Hempery

Ohotto Hemp
Original Sources
Pan World Traders
Rising Star Futon
Sharon's Finest
Stradom S.A.
Sue's Amazing Lip Stuff
Sunsports
SWIHTCO, The Swiss Hemp Trading Co.
Tomtex S.A.
Treefree Eco-Paper
Two Star Dog
U.S. Hemp
U.S. Textiles
Wise Up Reaction Wear

Manufacturer's Representative

Earth Pulp and Paper
Hemp Shak

Sourcing Consultants

Creative Expressions
Crop Circle Clothing
Green Machine
Hemp Cooperation
Hemp Educational Research Board
Hemp On-Line
Hempstead Company
Humboldt Industrial Hemp Supply Co.
Institute For Hemp
The Ohio Hempery

Original Sources
Pickering International
Ukrimpex

Contract Fabric & Garment Dyeing

ASA Aware Products
Humboldt Industrial Hemp Supply Co.
Mystic Eye

Publishers

Creative Expressions
Hemp Magazine
Hemptech
Hemp World Magazine
International Hemp Association
The Message Company
Perceptions
Solar Age Press
Sweetlight Books

Associations

Business Alliance for Commerce in Hemp
Hemp Educational Research Board
Institute For Hemp
International Hemp Association

Mail Order Catalogs

Earth Care
Natural Choice
Real Goods
Tomorrow's World

Writer

Hemp On-Line

Index by Product

Clothing

Greater Goods
Green Machine
Greener Alternatives
Green Underworld
Group W Bench
Hayward Hempery
Head Trips Hat Company
Hemcore, Ltd.
Hemp BC
The Hemp Club, Inc.
The Hemp Connection
Hempfully Yours
Hemp Head
Hemp Hop
Hemp is Hep
Hempstead Company
Hempstead Company Store
Hemp Textiles International
Hemptown
Hemp Works, Inc.
Hempy's
Humboldt Industrial Hemp Supply Company
Labyrinth Phassions and Costumes
Manashtash, Inc.
Mary Jane's Hemp Harvest
MAterials
Mindful Products
Montana Hemp Traders
Mystic Eye
Natural Rags and Hempery
The New Age Emporium
Of The Earth
The Ohio Hempery, Inc.

Ohotto Hemp
Pan World Traders
Pickering International
PlanetWise Products
Shakedown Street
Simply Better
Sow Much Hemp
Still Eagle Planetary
Texas Hemp Company
The Third Stone Hemp Products
Tomorrow's World
True North Hemp Company, Ltd.
Two Star Dog

Accessories
(includes handbags, wallets, belts, jewelry, patches, toys, etc.)

2000 BC
AdventureSmiths
AH Hemp
Alaska Green Goods
All Points East
American Hemp Mercantile
Artisan Weavers
ASA Aware Products
Australian Hemp Products
British Hemp Stores
Cannabest
Cannabis Clothes
Cannabiz
Canvas Hemp Company (CHC)
Chanvre en Ville

Community Market
Crop Circle Clothing
Crucial Creations
Cultural Repercussions
Deep See
Dharma Emporium
The Dolphin Song
Earth Care
Earth Goods USA, Inc.
Earth Wish
Earthware's, Inc.
Eastwinds Trading Company
Eco Goods
Eco-Wise
Ecological Wisdom
Ecolution
Emperor's Clothing Company
Everything Earthly
Exotic Gifts
First Hungarian Hemp Spinning Company
Forbidden Fruits
Greater Goods
Green Machine
Greener Alternatives
Green Underworld
Group W Bench
Hayward Hempery
Head Trips Hat Company
Hello Again
Hemcore, Ltd.
Hemp BC
The Hemp Club, Inc.
Hemp, Etc.

Hempfully Yours
Hemp Head
Hemp Sacks
Hempstead Company
Hempstead Company Store
Hemp Textiles International
Hemptown
Hemp Works, Inc.
Hempy's
Hip Hemp
Humboldt Industrial Hemp Supply Company
Lost Harvest
Manashtash, Inc.
Mary Jane's Hemp Harvest
MAterials
Montana Hemp Traders
Mystic Eye
Natural Rags and Hempery
The New Age Emporium
Of The Earth
The Ohio Hempery, Inc.
Ohotto Hemp
Pan World Traders
PlanetWise Products
Quick Distribution Company
Real Goods
Sexton Belt Company
Shakedown Street
Simply Better
Sow Much Hemp
Still Eagle Planetary
Sunsports
The Third Stone Hemp Products

Tomorrow's World
True North Hemp Company, Ltd.
Wise Up Reaction Wear

Hats

2000 BC
AdventureSmiths
AH Hemp
Alaska Green Goods
All Points East
American Hemp Mercantile
ASA Aware Products
Cannabest
Cannabis Clothes
Cannabiz
Chanvre en Ville
Community Market
Cultural Repercussions
Deep See
Dharma Emporium
The Dolphin Song
Earth Care
Earthware's, Inc.
Eastwinds Trading Company
Eco Goods
Eco-Wise
Ecological Wisdom
Emperor's Clothing Company
Everything Earthly
Exotic Gifts
Friendly Stranger
Greater Goods

Green Machine
Greener Alternatives
Green Underworld
Group W Bench
Hayward Hempery
Head Trips Hat Company
Hello Again
Hemcore, Ltd.
Hemp BC
The Hemp Club, Inc.
Hemp, Etc.
Hempfully Yours
Hemp Head
Hemp Hop
Hempstead Company
Hempstead Company Store
Hemptown
Hemp Works, Inc.
Hempy's
Hip Hemp
Home Grown Hats
Humboldt Industrial Hemp Supply Company
Labyrinth Phassions And Costumes
Lost Harvest
Mary Jane's Hemp Harvest
MAterials
Montana Hemp Traders
Mystic Eye
Natural Rags and Hempery
The New Age Emporium
Of The Earth
The Ohio Hempery, Inc.
Ohotto Hemp

Pan World Traders
PlanetWise Products
Quick Distribution Company
Shakedown Street
Sow Much Hemp
Still Eagle Planetary
Sunsports
The Third Stone Hemp Products
True North Hemp Company, Ltd.
Two Star Dog

Shoes and Sandals

2000 BC
Australian Hemp Products
Cannabest
Canvas Hemp Company
Chanvre en Ville
Crop Circle Clothing
Deep See
Dharma Emporium
Earthware's, Inc.
Eco Goods
Eco-Wise
Emperor's Clothing Company
Everything Earthly
Greater Goods
Green Machine
Group W Bench
Hemcore, Ltd.
Hemp BC
The Hemp Club, Inc.
Hempstead Company Store

Hemptown
Hemp Works, Inc.
Lost Harvest
Mary Jane's Hemp Harvest
Mindful Products
Mystic Eye
The Ohio Hempery, Inc.
Pickering International
Real Goods
Sow Much Hemp
Sunsports
The Third Stone Hemp Products
Tomorrow's World

Packs and Outdoor Gear
(includes nets, sails, marine products, canvas, etc.)

2000 BC
AdventureSmiths
AH Hemp
American Hemp Mercantile
Artisan Weavers
ASA Aware Products
Australian Hemp Products
Cannabest
Crop Circle Clothing
Earthware's, Inc.
Eastwinds Trading Company
Ecological Wisdom
Ecolution
Emperor's Clothing Company
Everything Earthly
Exotic Gifts

Exquisite Products Company
First Hungarian Hemp Spinning Company
Greater Goods
Green Machine
Green Underworld
Group W Bench
Hayward Hempery
Hemp BC
The Hemp Club, Inc.
Hemp, Etc.
Hempstead Company Store
Hemp Traders
Hemp Works, Inc.
Hempy's
Hip Hemp
Humboldt Industrial Hemp Supply Company
Lost Harvest
Magna International, Inc.
Manashtash, Inc.
MAterials
Montana Hemp Traders
Natural Choice
Of The Earth
The Ohio Hempery, Inc.
Ohotto Hemp
Pan World Traders
Shakedown Street
Simply Better
Stradom S.A.
Sow Much Hemp
True North Hemp Company, Ltd.
Two Star Dog

Fabric

2000 BC
AdventureSmiths
All Around The World Hemp
American Hemp Mercantile
ASA Aware Products
Australian Hemp Products
Cannabiz
Chanvre en Ville
Cotton and Willow
Crop Circle Clothing
Danforth International
Earth Pulp and Paper
Eastwinds Trading Company
Eco-Wise
Ecological Wisdom
Ecolution
Everything Earthly
Exotic Gifts
Exquisite Products Company
First Hungarian Hemp Spinning Company
Friendly Stranger
Green Machine
Greener Alternatives
Green Underworld
Hayward Hempery
Hemcore, Ltd.
Hemp BC
The Hemp Club, Inc.
Hempstead Company
Hempstead Company Store
Hemp Textiles International

Hemptown
Hemp Traders
Hemp Works, Inc.
Humboldt Industrial Hemp Supply Company
Makop Linen Mill
Magna International, Inc.
Mary Jane's Hemp Harvest
Montana Hemp Traders
Nagylaki Kenderfono Gyar Rt.
Naturetex International BV
Naturetex NYC
Of The Earth
The Ohio Hempery, Inc.
Pan World Traders
Pickering International
Stradom S.A.
Sow Much Hemp
Tomtex
Ukrimpex

Cordage
(string, rope, twine, yarns, and thread)

2000 BC
AH Hemp
Alaska Green Goods
American Hemp Mercantile
Australian Hemp Products
British Hemp Stores
Cannabest
Community Market
Crop Circle Clothing
Danforth International

The Dolphin Song
Earthware's, Inc.
Eco Goods
Eco-Wise
Ecological Wisdom
Ecolution
Emperor's Clothing Company
Everything Earthly
Exotic Gifts
Exquisite Products Company
First Hungarian Hemp Spinning Company
Friendly Stranger
Greater Goods
Green Machine
Greener Alternatives
Green Underworld
Group W Bench
Head Trips Hat Company
Hemcore, Ltd.
Hemp BC
The Hemp Club, Inc.
Hemp Head
Hemptown
Hemp Works, Inc.
Humboldt Industrial Hemp Supply Company
Magna International, Inc.
Mary Jane's Hemp Harvest
Nagylaki Kenderfono Gyar Rt.
Natural Rags and Hempery
The Ohio Hempery, Inc.
Pan World Traders
Simply Better
Sow Much Hemp

Still Eagle Planetary
The Third Stone Hemp Products
True North Hemp Company, Ltd.

Paper

2000 BC
AH Hemp
Alaska Green Goods
All Around the World Hemp
All Points East
Alma Rosa N.V. Belgium
American Hemp Mercantile
British Hemp Stores
Cannabest
CEL BT
Chanvre en Ville
Community Market
Danforth International
Dharma Emporium
Earth Care
Earth Pulp and Paper
Earth Wish
Earthware's, Inc.
Eco Goods
Eco-Wise
Ecological Wisdom
Ecolution
Ecosource Paper, Inc.
Emperor's Clothing Company
Evanescent Press
Everything Earthly
Exotic Gifts

Friendly Stranger
Greater Goods
Green Machine
Greener Alternatives
Green Underworld
Group W Bench
Hayward Hempery
Hemp BC
The Hemp Club, Inc.
Hemp Head
Hempstead Company Store
Hemptown
Hemp Works, Inc.
Humboldt Industrial Hemp Supply Company
Longevity Book Arts
Natural Rags and Hempery
The New Age Emporium
The Ohio Hempery, Inc.
Original Sources
Real Goods
Shakedown Street
Sow Much Hemp
Still Eagle Planetary
the Third Stone Hemp Products
Trefree Eco-Paper
True North Hemp Company, Ltd.

Body Care Products

2000 BC
AH Hemp
Alaska Green Goods
All Around The World Hemp
Alma Rosa N.V. Belgium
American Hemp Mercantile
Artha
Australian Hemp Products
British Hemp Stores
Cannabest
Chanvre en Ville
Community Market
Deep See
Dharma Emporium
Earth Care
Earth Wish
Eco Goods
Eco-Wise
Ecological Wisdom
Ecolution
Ecosource Paper, Inc.
Emperor's Clothing Company
Everything Earthly
Exotic Gifts
Forbidden Fruits
Green Machine
Greener Alternatives
Green Underworld
Group W Bench
Hayward Hempery
Hemp BC

The Hemp Club, Inc.
Hemp Essentials
Hemp Etc.
Hemp Head
Hemp Hemp Hooray
Hemp is Hep
Hempstead Company Store
Hemptown
Hemp Works, Inc.
Humboldt Industrial Hemp Supply Company
Mary Jane's Hemp Harvest
Montana Hemp Traders
Mystic Eye
Natural Rags and Hempery
The New Age Emporium
The Ohio Hempery, Inc.
Original Sources
PlanetWise Products
Quick Distribution Company
Shakedown Street
Sow Much Hemp
Still Eagle Planetary
Sue's Amazing Lip Stuff
The Third Stone Hemp Products
True North Hemp Company, Ltd.
Wise Up Reaction Wear

Home Accessories

(includes futons, sheets, pillows, towels, washcloths, etc.)

2000 BC
AH Hemp
Cannabest
Cotton and Willow
Earth Care
Earthware's, Inc.
Everything Earthly
Exotic Gifts
First Hungarian Hemp Spinning Company
Green Underworld
Hello Again
Hemp Essentials
Hemp Hemp Hooray
Hempstead Company
Hemp Works, Inc.
Humboldt Industrial Hemp Supply Company
Makop Linen Mill
Manashtash, Inc.
Mary Jane's Hemp Harvest
The New Age Emporium
Of The Earth
Ohotto Hemp

Foods

2000 BC
All Around the World Hemp
American Hemp Mercantile
The Boulder Hemp Company
British Hemp Stores
Cannabest
Community Market
Deep See
Dharma Emporium
Dr. Brady's Hemp Seed Delights
Earthware's, Inc.
Ecolution
Ecosource Paper, Inc.
Everything Earthly
Exotic Gifts
First Hungarian Hemp Spinning Company
Forbidden Fruits
Green Machine
Hayward Hempery
Hemp BC
The Hemp Club, Inc.
Hemp Essentials
Hemp Head
Hempstead Company Store
Hemptown
Hemp Works, Inc.
Herbal Products and Development
Humboldt Industrial Hemp Supply Company
Hungry Bear Hemp Foods
Mary Jane's Hemp Seed Snacks
Montana Hemp Traders

Natural Rags and Hempery
The New Age Emporium
The Ohio Hempery, Inc.
Original Sources
Sharon's Finest
Sow Much Hemp
Still Eagle Planetary
SWIHTCO The Swiss Hemp Trading Company
The Third Stone Hemp Products
True North Hemp Company, Ltd.
Ukrimpex

Agricultural Products
(includes birdseed, animal feed, bedding, compost, mulch, fertilizer, etc.)

Cannabest
Ecological Wisdom
Hayward Hempery
Hemcore, Ltd.
Natural Rags and Hempery
The Ohio Hempery, Inc.
Original Sources
PlanetWise Products
Ukrimpex

Building Materials

All Around the World Hemp
Exotic Gifts
Green Machine
Hemcore, Ltd.
Humboldt Industrial Hemp Supply Company

134

Nagylaki Kenderfono Gyar Rt.
Pan World Traders
Still Eagle Planetary
Ukrimpex

Industrial Products
(includes fuels, inks, paints, varnishes, etc.)

All Around the World Hemp
Australian Hemp Products
Eco-Wise
Exotic Gifts
First Hungarian Hemp Spinning company
Herbal Products and Development
Humboldt Industrial Hemp Supply Company
The Ohio Hempery, Inc.
Original Sources
Ukrimpex

Books, Magazines, Videos

2000 BC
American Hemp Mercantile
British Hemp Stores
Cannabest
Community Market
Creative Expressions
Deep See
Dharma Emporium
The Dolphin Song
Eco Goods
Eco-Wise

Ecological Wisdom
Everything Earthly
Friendly Stranger
Greater Goods
Greener Alternatives
Green Underworld
Group W Bench
Hemp BC
Hemp Head
Hempstead Company Store
Hemptown
Institute For Hemp
Mary Jane's Hemp Harvest
The Message Company
Montana Hemp Traders
Natural Rags and Hempery
The New Age Emporium
The Ohio Hempery, Inc.
Original Sources
Perceptions
Quick Distribution Company
Shakedown Street
Sow Much Hemp
Solar Age Press
Still Eagle Planetary
Sweetlight Books
Third Stone Hemp Products
True North Hemp Company, Ltd.

Addendum--New Listings Too Late to Classify

Aradia's Kitchen
Sarah Corning
342 Wythe St., Williamsburg, NY 11211
USA
718-388-2010
E-mail: aradia@amc.org
Distributor, Manufacturer.
Products: Cookies, cakes, bread, pies.

From the Ground Up
David Ross
PO Box 20251 Greeley Square Station, NY, NY
10001-0003
USA
212-564-0467   Fax: 212-971-0349
Importer
Full service fabric supplier/converter.
Products: Fabric, canvas, sourcing
consultants, import brokerage, contract fabric
and garment dyeing.

Grassroots of Hawaii
Helen Veldes
66-082 Kamehameha, Haleiwa, HI 96712
USA
808-637-6713   Fax: 808-638-1605
Retailer, Distributor, Manufacturer.
All clothes are made from hemp/silk.

Products: Men and women's clothing, handbags, wallets, accessories, hats, packs, fabric, rope, twine, printing paper, stationery, books, videos, jewelry.

Green Horizons
Larry Phillips
PO Box 5233, Burlington, VT 05402
USA
802-865-2646
Manufacturer.
Hemp seed foods, nutritious and delicious along with 26 flavors of hemp ice cream.
Products: Seeds (raw and roasted), food grade oil (bulk), salad oils, dressings, cookies, brownies, ice cream, muffins, pancake mix, hemp frozen yogurt, hemp flour, hemp baking mix, dog treats, animal nutrient supplements.

Green Monkey Creations
Stacy Mumm
PO Box 3352, Chico, CA 95927
USA
Toll-free: 800-436-7669 Fax: 916-892-2010
Manufacturer.
Products: Jewelry.

Harvest House
Angela Freese
2846 W. North Ave., Chicago, IL 60647
USA
312-292-1395 Fax: 312-292-9471
Toll-free: 800-975-HEMP

Distributor, Manufacturer.
Handmade by local artisans.
Products: Jewelry.

Hemp Heritage
Josephine Domsic
228 1/2 Thompson Blvd., Ventura, CA 93001
USA
805-641-3244
Manufacturer.
Lingerie made from hemp/silk for women
and sleep wear for men.
Products: Men and women's clothing.

Hemp Shak
Mark Hornaday
240 W. Foothill Blvd., Claremont, CA 91711
USA
909-398-1041   Fax: 909-398-1041
Retailer.
Southern California's exclusively hemp store.
Also a Manufacturer's Representative.
Products: Men and women's clothing,
handbags, wallets, accessories, hats, packs,
outdoor gear, shoes, wash cloths, fabric, rope,
twine, stationery, printing paper, soaps,
massage and body oils, salves, lip balms,
seeds (raw and roasted), food grade oil (bulk),
jewelry, books, videos.

Hemp Sprouts
Michelle Carter
PO Box 2629 Scottsdale, AZ 85252-2629
USA
602-675-0709   Fax: 602-675-0709
Retailer, Distributor.
Products: Baby cloths, accessories, children's sheets, stuffed toys.

Hempire
Yolanda Ledesma
PO Box 1114, Santa Barbara, CA 93102
USA
805-967-2032
Manufacturer, Wholesaler, Designer.
Clothing for the soul.
Products: Men, women and children's clothing.

Herban Home
David Marcus
833 Manning Ave., Toronto, Ontario M6G 2W9
Canada
416-535-3497
Distributor, Manufacturer.
New and innovative hemp home furnishing.
Products: Tablecloths, napkins, placemats, blinds, lamps.

Acknowledgments

Thanks to the following people and organizations for contributing information used in this Directory:

Institute of Hemp: *World History of Commercial Hemp*
Hemptech: *The Earth's Premier Renewable Resource*
Earth Pulp and Paper: *Hemp Paper*
Herbal Products and Development: *The Rediscovery of Hemp Seed Oil*

Thanks to the following companies for providing samples for front cover photography:

Hempy's
Ecosource Paper
Hemp Essentials
Longevity Book Arts
Artha
Evanescent Press

ORDER FORM

Phone, fax or write if you would like a copy of our catalog which lists all books, videos, etc. that we produce and distribute.

Telephone orders: Call 505-474-0998
   to charge to Visa, MC, AE or Discover
Fax orders: 505-471-2584

Mail orders: The Message Company
4 Camino Azul
Santa Fe, NM 87505

CODE #	AUTHOR	TITLE	QTY	PRICE	TOTAL
D20268	Willis Harman, PhD	Spirituality in Business: The Tough Questions Color video, 60 minutes	____	19.95	____
D20276	Richard Barrett	Unfolding of the World Bank Spiritual Unfoldment Society Color video, 60 minutes	____	19.95	____
D20284	Sharon Morgen	Exploring the New Paradigm in Sales Color Video, 120 minutes	____	29.95	____
D20292		Buying Facilitation: The Service that Sells Color video, 60 minutes	____	19.95	____
D20306	Martin Rutte	Livelihood: Growing Spirit at Work Color video, 60 minutes	____	19.95	____
D20314	Judith Thompson, PhD	Moving from Corporate Social Responsibility to Corporate Spirituality Color video, 60 minutes	____	19.95	____
D20322	Michael Horst and Brooke Warrick	Spirituality in Real Estate Color video, 60 minutes	____	19.95	____
D20330	Hope Mineo	Harnessing the Power of Business to Create a World Worth Having Color video, 120 minutes	____	29.95	____
D20349	Mary Ruwart, PhD	Why Competition Always Starts with Cooperation Color video, 60 minutes	____	19.95	____
D20357	Hal Brill	The Money Seminar: Investing from the Heart Color video, 60 minutes	____	19.95	____
D20365	Robert Ellis	Hypertext and the Holy Ghost: Cyberspace, Telecommuting, The Internet, Virtual Reality: The New Frontier Color video, 60 minutes	____	19.95	____
D20373	Ted Nicholas	Why 97% of our Conventional Wisdom is Wrong and How to Change it Color video, 120 minutes	____	29.95	____
D20381		Success in Life is Found in Balance Color video, 60 minutes	____	19.95	____
D20403		The Crucial Difference Between Positive and Negative Selfishness Color video, 60 minutes	____	19.95	____
D20411		Examining 17 Areas of your Belief System Color video, 120 minutes	____	29.95	____
Complete Set of 15 videos		(if purchased individually: $339.25)	____	229.95	____

Subtotal: ____
USA shipping: <u>2.00</u>
Extra shipping: ____
NM residents add 5.75% sales tax: ____
Total: ____

Extra shipping charges:
- Priority mail: add $3.00
- AK, HI, Canada: add 10%
- All other countries: add 25%
- Foreign Air: add $8.00 per tape

❏ Enclosed is a check/money order for total
❏ Please charge to my   ❏ Visa   ❏ MC   ❏ AE   ❏ Discover

Card # _____ Expiration Date: _____

Signature _____

Name _____ Address _____

City/Sate/Province_____Zip/Postal Code _____

Phone: _____

NEW HANOVER COUNTY PUBLIC LIB.
3 4200 00448 7480

Phone, fax or wr ... catalog listing all books, videos, etc. that we produce and distribute.

Telephone Orders: Call 505-474-0998 to charge your VISA, MasterCard, American Express or Discover

Fax Orders: 505-471-2584

Mail Orders: The Message Company
4 Camino Azul
Santa Fe, NM 87505

CODE	TITLE	QUANTITY	PRICE	TOTAL
D20055	**International Directory of Hemp Products and Suppliers** 144 pages, 6 x 9, paperback.	______	x $29.95 =	______
D20128	**How to Find the Best Lawyers . . . and Save 50% on Legal Fees** 200 pages, $5^1/_2$ x $8^1/_2$, paper back.	______	x $14.95 =	______
D20101	**History of the American Constitutional or Common Law** 144 pages, $8^1/_2$ x 11, paperback.	______	x $11.95 =	______
D20020	**The Physics of Love** 156 pages, $8^1/_2$ x 11, paperback.	______	x $15.95 =	______
D20063	**Spiritual Vampires: The Use and Misuse of Spiritual Power** 256 pages, 6 x 9, paperback.	______	x $14.95 =	______
D2008X	**Nikola Tesla's Earthquake Machine** 176 pages, $8^1/_2$ x 11, paperback.	______	x $16.95 =	______
D20039	**Universal Laws Never Before Revealed: Keely's Secrets** 288 pages, $8^1/_2$ x 11, paperback.	______	x $19.95 =	______
D20047	**Grazing Through the Woods with the Herb Man** Color video.	______	x $19.95 =	______

SUBTOTAL ______
PPING $2.00
IPPING ______
5.75% ______
TAL $______

Extra Shippi
- Priority Ma
- AK, HI, Ca
- All other co
- Foreign Ai

❑ Enclosed is
❑ Please charg
[ ] Americ

CARD # ______

EXP. DATE ______

SIGNATURE ______

8/97

NEW HANOVER COUNTY PUBLIC LIBRARY
201 Chestnut Street
Wilmington, N.C. 28401

GAYLORD S

ML